麹本

KOJI for LIFE

なかじ 著　BY NAKAJI

KOJI for LIFE
麹本
NAKAJI
なかじ

はじめに …2

1.麹とは？ …4

2.麹をつくる …10

  道具 …10

  タイムスケジュール …14

  米を蒸す …16

    ①洗米 …17

    ②浸漬 …18

    ③ザル上げ …20

    ④蒸し …22

    ⑤蒸しあがり …24

  種麹をつける …26

    ⑥蒸し取り …26

    ⑦種切り …28

    ⑧包み込み〈発芽期〉…30

  菌糸を成長させる …32

    ⑨盛り〈菌糸成長期〉…32

    ⑩仲仕事 …34

    ⑪仕舞仕事 …35

    ⑫最高温度をチェック〈酵素蓄積期〉…36

  できあがり …37

    ⑬完成 …37

    ⑭出麹 …38

    ⑮枯らし …40

    麹菌の生態 〜原理を知る …42

3.麹でつくる …45

  甘酒 …46

  乳酸発酵甘酒 …47

  味噌 …48

  菩提酛／どぶろく …50

  塩麹 …53

  麹パウダー …54

番外編 もうひとつの麹つくり …55

  田んぼの稲麹から育てる麹 …55

  麹からの種を継ぐ 〜種麹をつくる …58

種麹の入手先一覧 …61

おわりに …62

Introduction ... 2

1. What is Koji? ...4

2. Making Koji ...10

  Tools and Equipment ...10

  Timings ...14

  Steaming rice... 16

    1. Senmai ...17

    2. Shinseki ...18

    3. Zaruage ...20

    4. Mushi ...22

    5. Mushiagari ...24

  Inoculate with koji spores ...26

    6. Mushitori ...26

    7. Tanekiri ...28

    8. Tsutsumikomi<Germination phase> ...30

  Mycelial growth ...32

    9. Mori <Mycelial growth phase> ...32

    10. Naka ...34

    11. Shimai ...35

    12. Checking maximum temperature   <Enzyme production phase> ...36

  Completed rice koji ...37

    13. Completed rice koji ...37

    14. Dekoji ...38

    15. Karashi ...40

    Basic koji biology ...42

3. Using koji in your everyday cooking ...45

  Amazake ...46

  Lacto-fermented amazake ...47

  Miso ...48

  Bodaimoto/Doburoku ...50

  Shio koji ...53

  Koji powder ...54

Appendix – Koji making – Another story ...55

  Cultivating koji spores from inakoji ...56

  Tanetsugi ~Propagating another generation of koji   spores ...59

List of koji spore producers ...61

Afterword ...62

※ページ内の QR コードを読み取ると、その作業の動画サイトが開きます。

INTRODUCTION

はじめに

　　暮らしの中で、麹を育てる。

　　この本は、そのための方法を書いた本です。

　　こんにちは。なかじです。この本を手に取っていただきありがとうございます。ぼくは「麹の学校」という麹つくりや発酵を学ぶための学校を運営したり、世界中で実際に麹つくりを教える活動を行なったりしています。もともとは日本酒の造り酒屋でお酒をつくっていた「蔵人」です。蔵人とは日本酒をつくる技術者のことです。

　　7〜8年ほど前から、それまで特殊な専門家の技術や知識であった「麹つくり」について、一般の人から学びたいという声が聞こえ始めました。初めは友達に

「麹つくり、教えてよ」

　　といわれたのがきっかけで2泊3日のワークショップでやってみました（麹つくりは3日間かかるから）。そこでの最初の参加者が「わたしの町でもやってくれませんか？」と次の町に呼ばれて、またその次の参加者に呼ばれて…と、そんな感じで「麹つくりワークショップ」は自然に全国に広がっていきました。今では世界中に麹をつくる仲間がいます。もともと日本の農村で麹つくりは行なわれていましたが、都会ではその限りではなく、その頃から麹つくりがだんだんと一般の人にも広がっていきました。

　　ワークショップを開く中で、本来、蔵の整った環境でやっていた「麹つくり」を、普通に生活している家の台所でやるためにはどうしたらいいか、その試行錯誤を繰り返し、田舎でも都会のマンションでも、安定して、安全に、確実につくれることを目指し、「蔵の中のやり方」から「暮らしの中でのやり方」に変わっていきました。

　　その過程は、これまでのワークショップの参加者や麹の学校オンラインスクールのメンバー、そして世界中の講師の仲間、これまで関わってくれたすべての人々の経験・発見・気づきの積み重ねです（ありがとう！）。今回はその**暮らしの中で安全に確実に、世界中どこでも手に入る道具で麹を育てる**という方法をお伝えしたいと思います。麹は、環境を用意すれば世界中でつくることができます。

Make koji part of your everyday life.

Hello! My name is Nakaji. Thank you for choosing my book.

Let me tell you a little about myself before explaining why I wrote this book. I run a school in Japan called "The Koji Academy" where students learn about making koji and fermentation. I also travel around the world, sharing my knowledge of koji with people in other countries.

I started off working at a sake brewery. Sake brewing requires specialist skills, which I acquired through many years of experience. Making koji was also thought to require professional skill and technique, and so limited to production by trained craftsmen. Despite that, interest in making koji has increased over the last seven or eight years. My journey began when I hosted a three-day workshop for some close friends who asked me to teach them how to make koji. One of the attendees asked me to host the same workshop in their town. When I did that, another attendee asked for another workshop. Before I knew it, I was holding workshops all over the country, and I now have koji friends all over the world. Koji was traditionally made in rural areas of Japan, but this surge of interest has resulted in more people making koji for themselves in both the countryside and the cities.

When I worked in a sake brewery, I made koji in a carefully controlled environment. Some elements of this process were difficult to replicate at home, so I looked for more accessible methods that didn't compromise stability or safety. This search led to a gradual transformation in how I make koji.

All this was made possible by the people I met on my koji-making journey: workshop attendees, students at the Koji Academy, and fellow koji teachers. I am truly grateful to everyone who generously shared their experiences, discoveries, and inspiration.

In this book, I want to share the koji-making method I have developed. I believe it is **the best and most accessible method for making koji stably and safely, using equipment readily available anywhere in the world**. I hope to show you how making koji is not limited by your environment, rather the key is to prepare the optimum environment for koji to grow.

CHAPTER 1

What is

"Koji" refers to koji microbes grown on steamed grains such as rice, soybeans or barley. It is essential for the fermentation processes used to make sake, soy sauce and miso.

麹が日本の食文化をつくってきた

　「麹・糀」（こうじ／KOJI）って知っていますか？

　麹とは「ニホンコウジカビ」とも呼ばれ、人間の生活圏で古くなったパンやご飯にふわふわと生えてくる赤や黒や白のカビの仲間です（ニホンコウジカビは黄緑色です）。カビの食文化は世界中にあります。ヨーロッパでは乳製品によく使われ、アジアでは穀物の発酵に使われます。その中でもニホンコウジカビは日本人の長い歴史をともに歩み、日本の食文化の中で培われてきたカビです。

　ほとんどの日本人は意識することはないと思いますが、もし麹が日本になかったら、現在の日本の食文化は全然違うものになっていたはずです。麹がなければ日本酒がない、味噌もない、醤油もない。今の和食が成立しません。実は和食文化の根底を支える基本調味料は、麹から生まれる発酵食品だからです。

　"和食の基本調味料 さしすせそ"は、調理に使う順番とされ、その順番と調味料はこんな感じです。

さ＝酒or砂糖　し＝塩　す＝酢　せ＝醤油　そ＝味噌

　砂糖と塩は世界中にあります。その二つ以外の、日本独特の風味、いわゆる「日本っぽさ」を生み出す酒、酢、醤油、味噌は、麹から生まれる発酵調味料です。長期で海外旅行に行くと旅の終盤、無性に醤油味のものを欲してしまうのは、多くの日本人に経験があると思います。日本人はある意味、「醤油中毒」とも言えます。醤油がなければ日本人としてのアイデンティティを保てません。少なくともぼくはそうです。

KOJI SHAPED THE DEVELOPMENT OF JAPANESE FOOD CULTURE

Koji is often yellow or green. It is related to species of mould you see growing on stale bread or rice, which are usually red, black or white. Using mould to ferment food isn't new or particularly unusual. Mould has been widely used in Asia to ferment cereals (e.g. tempeh, jiang, furu), and to produce cheese in the West. Koji has a long history in Japan and is a major driver in the development of its food culture. Japanese food would be unrecognisable without koji – there would be no sake, soy sauce, miso or rice vinegar. All essential Japanese condiments, the flavours that characterise Japanese cuisine, are made by fermentation methods that use koji. When Japanese people travel, many start craving these flavours of "home", exposing their intimate connection to koji. The process of making soy sauce starts with inoculating a mixture of roasted and crushed wheat and steamed soybeans with koji spores. The grains and koji are mixed with brine and left to ferment, and the liquid squeezed from this mixture is soy sauce. Making miso starts by steaming and crushing soybeans, which are mixed with koji and salt and left to ferment.

Sake is made from rice and koji, and deeply intertwined with the concepts of Japanese religion, which in turn led to the development of traditional performing arts and music. So you could say that the whole of Japanese culture has been nurtured by koji.

Koji

麹とは？

麹は蒸した米や大豆、麦などに
麹菌を繁殖させた菌の集合体。
酒や醤油、味噌などの和食の基本調味料を
つくる際に欠かせないスターターです。
麹菌は日本の食文化を育ててきた
お婆ちゃんのような存在です。

ちなみに醤油は、麹菌を炒った麦と蒸した大豆に培養して、その後、塩と水で合わせて数ヵ月間おいて発酵・熟成させ、それをしぼった液体です。味噌は、蒸した大豆をつぶして麹と塩を混ぜて、醤油同様に発酵・熟成させたもの。

さらに酒は広く芸能、音楽、宗教に深く関わり、多くの日本の文化を育んできました。麹は日本文化を育てたお婆ちゃん〈grandma／グランマ〉のような存在です。

麹菌の生態を生かした醸造

ニホンコウジカビの学名は「*Aspergillus oryzae*／アスペルギルス・オリゼー」です。アスペルギルス属は世界中に広く分布し、毒性を持つものが多いので、世界的には警戒されることが多いカビです。オリゼーは、その中では毒性を持たないカビの一つです。

硬い殻を持った胞子を飛ばして、栄養と水分のあるところに付着すると発芽し、根っこである菌糸を伸ばして成長します。成長すると、空気中に分生子柄と呼ばれる「毛」を伸ばし、その先端に丸い球体（頂のう）をつくります。その表面に胞子（分生子）を数珠なりにつくり、これが次世代の種になります。麹の胞子は硬い殻に包まれ、乾燥させると1年以上もその性質を保つことができます。だから麹菌は、胞子の形でその性質を維持し、保存と移動ができるのです。

この種（麹菌の胞子）を培養して製品化し、麹菌の種を売る専門の業者が日本では500年以上前から存在し、産業として発達しました。この業者を「種麹屋」といい、古くは「もやし屋」と呼んでいました。種麹屋さんのおかげで日本では、質の高い純粋な種麹を安定して手に入れることのできる環境

EXPLOITING KOJI'S BIOLOGY FOR FERMENTATION

The scientific name for koji is Aspergillus oryzae. It is a member of the Aspergillus family of moulds and is found all over the world. Other Aspergillus moulds produce toxins, but the domesticated Aspergillus oryzae does not. Koji releases hard-shelled spores which, when exposed to optimum moisture and nutrition, germinate and start to grow roots (mycelia). The growing koji then extends hairs (conidiophores) with small spherical structures (sporangiospores) on their tips. Spores like beads on a string are produced on the surface, the seeds of the next generation. As koji spores are protected by a hard shell, they remain viable for over a year when dried properly. This means that koji is easily preserved and traded in spore form.

Moyashi-ya, specialist companies that culture and sell koji spores, have existed for over 500 years in Japan. Their presence has ensured a supply of high quality, pure koji spores and contributed to the development of Japanese fermentation and brewing. Today there are over 3,000 companies making fermented food and drink such as sake, miso, soy sauce, rice vinegar, and mirin. They all rely on a very small number of koji spore producers - only ten are registered with the koji trade association.

Once grown, koji produces a range of enzymes from the tips of its mycelia. The enzymes break the starch and protein in grains down into sugar and amino acids, and koji feeds on these smaller

が整い、発酵醸造産業が発展しました。今でも日本の酒・味噌・醤油・酢・みりんなどの3000〜4000社の発酵醸造メーカーは、わずか10軒ほどの種麹屋さんから麹の種を仕入れています。

　そして、麹菌は成長するときに菌糸の先端から多種多様な酵素を生産します。この酵素が、穀物のでんぷんやたんぱく質を、糖やアミノ酸に分解します。麹菌はこれらを栄養にします。本来、この酵素は麹菌自身が生育するために生産するのですが、実はこの麹菌の酵素を利用してつくるのが酒、味噌、醤油、みりんなどの発酵食品です。麹菌の生態を上手に生かし、互いに影響を受けながら、変化・発展してきたのが、日本の麹発酵文化なのです。

麹つくりは酵素つくり

　一般的に「麹」（KOJI）と使われるときは、蒸した米や大豆、麦などの穀物に麹菌（ニホンコウジカビ・A. oryzae）を培養させたものを指します。米麹や麦麹、豆麹、醤油麹（麦と大豆）などがあります。酒や醤油、味噌などをつくる際に欠かせない発酵を始めるためのスターターです。では、麹の何が酒、味噌、醤油を美味しくしてくれるのか？というと、それは麹のつくり出す「酵素」です。

　麹菌は穀物を培地とし、発芽して菌糸を伸ばして、菌糸の先から酵素を出し、酵素で培地を分解して栄養にし、成長していきます。米麹の場合、蒸し米を培地として、麹菌にとって最適な温度や湿度に整えることで、菌糸が蒸し米内部によく伸び、酵素が多く生産されます。

　米麹のできあがりの姿というと、表面に花を咲かせたよう

molecules. So although koji produces these enzymes for itself, they are invaluable for us to produce sake, miso, soy sauce, and mirin. Koji's biology has been instrumental in the development of Japan's fermentation culture.

MAKING KOJI MEANS MAKING ENZYMES

"Koji" generally refers to steamed grains such as rice, soybeans, or barley, with koji grown on them. Names can also be based on the grain or final product: rice koji, barley koji, soybean koji or soy sauce koji (mixture of wheat and soybeans). These inoculated grains are essential to start the fermentation processes for sake, soy sauce and miso. So how does koji make sake, miso, and soy sauce so tasty? The answer lies in the magical power of enzymes.
As mentioned, koji mould grows on grains. After germinating, it grows mycelia that produce enzymes. These enzymes break down the grain, and koji digests the resulting molecules to keep growing. Take rice koji as an example. When given optimum temperature and moisture, koji grows mycelia around and into steamed rice and produces plenty of enzymes.
People tend to image a visually attractive, fluffy microscopic "flower" as the final form of rice koji, and often aim for something that looks like that when making koji for the first time. However, although fluffy koji is very appealing and fun to look at, appearance does not predict success when making koji. The real goal is to produce the highest quantity of

なふわふわとした毛が伸び、カビが生えた状態を思い浮かべます。初めて麹をつくるときは、ついこの見た目の状態を目指してしまいます。でも麹つくりは、その見た目にすることが目的ではなく、麹菌が菌糸を伸ばすときに生産する「酵素」を得るのが本当の目的です。厳密に言えば麹が味噌や醤油を美味しくするのではなく、麹の「酵素」が原料を分解し、甘みや旨みをつくり、味噌や醤油を美味しくし醸してくれるのです。

　麹つくりとは、培地である米や麦に「麹菌を育て酵素を生産してもらう」こと。麹のふわふわの見た目がゴールではありません（それももちろん麹つくりの魅力の一つではありますよね）。目的は、そのさらに先、麹を使って美味しい味噌や醤油、酒をつくることです。そして、それを人生の喜びとして「あぁ、美味しい〜」と味わうためです。少なくとも日本の麹文化の中では「より美味しい味噌や酒をつくるためには、どうやって麹を育てればいいのか？」を目的に麹つくりの技術が発展してきました。

　必要なのは、麹菌を成長させるということではなく、酵素が充分につくられることなのです。まず目的を、「麹菌を育てる」から、「麹菌に酵素を多く生産してもらうにはどうすればいいのか？」という視点に移すことが大切です。

　麹つくりは酵素つくりなのです。

酵素が発酵食品の味を決める

　さて、では酵素とはなんでしょう？
　ヒトでいうと唾液、胃液、膵液などの消化液が、ヒトがつくる酵素です。ヒトを含むすべての生物の生命活動に関わ

enzymes possible in the mycelia. In fact, it is not koji that makes miso or soy sauce so tasty. Their flavour comes from koji enzymes that decompose the grains and produce sweetness and umami. So, the true goal when making koji is making tasty condiments that add flavour and joy to life. Throughout koji's long history, techniques have been developed to further improve the taste of koji-based condiments. Fermenters have continuously pushed the envelope when making koji to reach their goals. So what you need to make good miso, sake and soy sauce is not well-developed koji, but rather well-balanced, enzyme-rich koji. You must shift your focus from simply making koji to making koji in an environment that encourages production of these essential enzymes. This is why I stress that **the purpose of making koji is to make enzymes.**

THE ROLE OF ENZYMES IN THE FLAVOUR OF FERMENTED FOODS

What are enzymes?
Digestive fluids, such as saliva and gastric or pancreatic fluids, contain the enzymes that make it possible for living creatures to survive. The most important enzymes when making koji or fermenting are hydrolases, which break down other molecules in the presence of water. Koji produces a treasure trove of enzymes. Approximately 30–100 of them are used in industrial applications, but it is thought to produce even more. New koji enzymes are still being discovered so the precise number is as

り、種類はたくさんありますが、麹つくりや醸造においての酵素とは、水分が豊富にある環境下で、対象物を分解する「加水分解酵素」です。

　麹は酵素の宝庫とも呼ばれ、たくさんの種類の酵素を生産します。その中から約30〜100種類ほどが産業的に使われ、潜在的にはそれ以上に多くの酵素があるとされます。研究とともに新たに発見されて数が増えることもあるので、正確な数はわかりません。

　その中で主に発酵醸造に関係する二つの系統があります。それが、アミラーゼ系とプロテアーゼ系です。

　アミラーゼは「でんぷん分解酵素」とも呼ばれ、でんぷん

yet unknown. Of the known enzymes, two types are key to fermentation and brewing: amylases, which break starch into sugar (polysaccharide or monosaccharide); and proteases, which break proteins into peptides or amino acids. Amylases produce sweetness, while proteases create umami.

I cannot stress enough that the quantity of enzymes in koji is key to the flavour of fermented products. Big, bold flavours cannot be produced by small quantities of enzymes. The balance between amylase and protease is also important, as this determines the final balance of flavours (i.e. predominantly sweet or umami). Amylase is most important when making amazake because you need to break the starch in rice down into sugars. On the other hand, protease is more important when making soy sauce because you want to break down the protein in the soybeans. Miso has a good balance of sweetness and umami, so a good balance of amylase and protease is crucial. The main enzyme you need depends on the fermented product and you must adjust the balance accordingly.

WHY MAKE KOJI YOURSELF?

Koji-based condiments can be made with koji bought from local koji shops or fermented food companies.

Sun
light
A. Oryzae
CO₂
Water H₂O
Oryza. Sativa
rice
C₆H₁₀O₅
C₂H₄O₂
Vinegar
CO₂
koji
CO₂
C₂H₆O
cooked rice
C₂H₁₂O₆
Sake
Amazake

What is Koji

を分解して、糖（多糖類や単糖）にします。

プロテアーゼは「たんぱく質分解酵素」と呼ばれ、たんぱく質を分解して、ペプチドやアミノ酸にします。

味としてはアミラーゼが甘みを、プロテアーゼが旨みをつくります。

自分で麹をつくる意味

味噌や醤油をつくるのに、麹づくりのプロである地域の麹屋さんや、醸造蔵の麹を使うことは、できあがりの食品の味を安定させ、手づくりのハードルを下げてくれます。そしてそれは地域の食文化を支え、次世代に伝えることにつながります。ですから、もちろん市販の麹であっても、発酵食品をつくるのはすばらしいことだと思います。

では、なぜ自分で麹をつくるのか。麹を自分でつくると、できあがりの麹の酵素の量、またアミラーゼとプロテアーゼのバランスを自分で自由に決めることができるのです。つまり、その発酵食品の味を自分で自由につくれるということです。そして、その過程では微生物にふれる楽しさ、蓋を開けた瞬間の喜びなど、手づくりだからこそ感じられる面白さがあり、みんな麹つくりにはまるのです。それには技術と経験が必要ですが……ガンバって！この本でちゃんとお伝えします。そして、このチャレンジと、少しずつ思い通りの麹がつくれるようになる道のり（そしてできあがったときの麹の愛おしさ！）が麹を自分でつくる醍醐味です。

さぁ、それでは麹をつくってみましょう！

The koji they produce is consistent in quality, and of course buying it is much easier than making koji yourself. Buying from local businesses supports local food culture and ensures it is passed down to the next generation. So buying koji is good for both you and the local community.

So, what are the benefits of making koji yourself? For me, making your own koji gives you freedom. You can choose the quantity of enzymes and control their balance, plan the flavours of the final product, and flexibly adjust them to respond to demands or preferences. On top of this, you get to feel the presence of microorganisms and interact with them during the fermentation process. This all adds to the excitement of opening the lid of a container of homemade condiments after a long wait, and the amount of satisfaction you feel when savouring your cooking. These are just a few of the many reasons why people fall in love with making koji. There is a learning curve and it requires lots of patience, skill and experience. But don't worry, that's what this book is for. The excitement you feel when faced with a new challenge, the gradual increase in confidence that comes with practice, and the satisfying sense of achievement are all part of the joy of making koji.

So let's make koji!

CHAPTER 2

Making

The method of making koji described in this book is one I have modified for making smaller batches. It uses readily available tools and equipment so you can make koji anywhere using local rice. I have used my experience making koji all over the world to refine this method, and will continue to do so.

Koji

麹をつくる

この本で紹介するのは、世界中で手に入る道具、
その環境、そこの米を使い、マンションでも田舎の家でも、
最小単位でつくれるように工夫した方法の一つです。
実際にぼくが世界各地を訪れながら試してみて、
改善を繰り返しながらこの方法になりました。
そして、もちろんこれからも進化していきます。

TOOLS AND EQUIPMENT 道具

今回の麹つくりの道具は、世界中どこでも手に入るものを選択しています。種麹以外は特別なものは使いません。どこでも手に入るシンプルな道具にデジタルな機器、両方を活用して誰でも世界中で麹がつくれます。

As mentioned, all the tools and equipment are easily to find. There is nothing special to prepare apart from the koji spores. Some simple electrical equipment is all you need to make koji.

① うるち白米
② 種麹
③ 茶こし
④ 蒸し布・さらし布
⑤ 温度計
⑥ はかり
⑦ スケッパー
⑧ 蒸し器
⑨ ステンレスバット
⑩ 電気毛布

1. Rice
2. Koji spores
3. Tea strainer
4. Loose-woven cotton/linen cloth (e.g. tea towel) and light cotton/linen cloth (e.g. muslin)
5. Thermometer
6. Scale
7. Dough scraper
8. Steamer
9. Stainless steel trays
10. Warming equipment (e.g. electric blanket)

①うるち白米…300g

粘りの少ないタイプの白米がオススメです。もち米のような粘る米は米粒どうしがくっつき、麹菌が菌糸を伸ばしにくくなります。保存状態がよければ古米もいいでしょう。

②種麹（米の0.1〜0.3%重量）…1g

種麹は、麹菌の胞子を集めた種のこと。粉状と粒状があります。色は緑や白。麹つくりに使用する標準量は、生米1kgに対し種麹1g、つまり生米の0.1%重量の割合ですが、家庭で少量つくる場合は生米の0.1〜0.3%重量が目安。今回は生米300gに対して種麹は1gにします。少量の米には、種麹は少し多めのほうがまんべんなく振れます。用途別に甘酒用や酒用、味噌用、醤油用、原料別に米、麦、大豆などの種類があります。全国の種麹屋やインターネット通販などで買えます。（p61の種麹の入手先一覧参照）

③茶こし

少量の種麹を均一に振り落とすことができます。種麹を木綿の布で包んで振ってもいいです。

④蒸し布…1枚

目の粗い木綿の布で、蒸気をよく通します。米を蒸すときに蒸し器に敷き、米を包むのに使います。

④さらし布…1枚

薄手の木綿の布。蒸し器に敷いたり、麹つくりの途中のバットにかぶせ、湿度を与えたりします。

⑤温度計

蒸し米の温度を測ります。p10の写真のようなコードの先に温度センサーがついた、室温と米の温度が両方測れるデジタルのものが便利。

1. White non-glutinous rice...300g

Non-glutinous rice is better as koji cannot grow its mycelia on sticky or glutinous rice. Any rice can be used if it has been stored properly.

2. Koji spores (0.1-0.3% of uncooked rice weight)...1g

Koji spores are harvested koji seeds. They come in two different forms, powder and grains, and are usually green or white. In general, 1g of spores is used for 1kg of uncooked rice (0.1% of rice weight). However, for smaller quantities you can use 0.1-0.3% of the rice weight. In this book, I use 1g koji spores for 300g of uncooked rice (around 0.3%). It's easier to use a little extra for small quantities, as it helps to sprinkle the spores evenly. Some spores are for making specific products (amazake, sake, soy sauce etc.), while others are specific to a substrate (rice, barley, soybeans etc.). You can buy koji spores directly from koji spore producers or online shops (please refer to the list of koji spore producers on p.61).

3. Tea strainer

Use a tea strainer to sprinkle the spores evenly. Alternatively, you can wrap the spores in a thin cotton/linen cloth and shake the bundle over the steamed rice.

4. Loose-woven cotton/linen cloth (e.g. tea towel)

Used when steaming rice. Choose loose-woven, lint-free cotton or linen (e.g. a tea towel) to allow steam to pass through. This cloth is placed in the steamer and folded over to wrap the rice.

4. Light cotton/linen cloth (e.g. muslin)

Choose light, lint-free cotton or linen (e.g. muslin). This cloth has many uses: it can be placed at the bottom of the steamer to absorb excess water, or used to cover stainless steel trays and provide moisture.

5. Thermometer

Used to measure the temperature of the steamed rice. A thermometer with a probe is ideal. The thermometer shown on p.10 can measure both room temperature and rice temperature.

⑥はかり

種麹の重さを量ります。0.01g
以下が量れるデジタルのものが
いいです。

種麹は米を蒸す間に量る。受け皿
の上に茶こしをおいて量るとよい。

Weigh the spores while the rice is steaming.
I recommend placing the tea strainer on the
scale before weighing out the spores.

⑦スケッパー（またはスプーン）

蒸し米を取り出し、混ぜたり広げたりするときに使います。

⑧蒸し器

　麹つくりに使う原料は、主に蒸した穀物です。水の中
で加熱する、つまり炊飯した米は、培地としては水分が
多すぎて麹菌が成長しにくく、雑菌が繁殖しやすくなり、
麹つくりには向きません。蒸す道具は必ず用意してくださ
い。蒸し器の材質には木製と金属製があります。理想は
和せいろ（写真右）や中華せいろ（写真左）などの木製
ですが、なければ金属製（写真奥）でもいいでしょう。

　蒸し器がなければ、寸胴鍋にザルをはめて中空に浮か
し、ザルに米を入れ、ザルの下に直接触れない程度の
湯を沸かして蒸し器の代わりにします。それもなければ、
簡単な日曜大工のできる人ならすぐに蒸し器の代わりをつ
くれます。仕組みは簡単。寸胴
鍋の上を全部カバーできるくら
いの面積の木の板を用意し、そ
の真ん中に穴（3〜4cm）を開け
て、上に前後左右の四面を囲っ
た木の箱枠をおき、中には米を
中空でおけるような工夫をすれ
ばよいのです。ザルをそのまま
おいてもいいでしょう。米は蒸
し布で包み、蒸気が逃げないよ
うに蓋をして蒸します。

6. Scale

Used to measure the weight of the spores. Choose a
digital scale accurate to one hundredth of a gram (0.01g).

7. Dough scraper (or spoon)

Used to take the steamed rice out of the steamer and
spread it onto the tray. Also used to mix the rice when
sprinkling the spores.

8. Steamer

Grains must be steamed before being inoculated with
koji. Boiled rice is not suitable for making koji as it
contains too much moisture, which makes it difficult
for koji to grow. Boiled rice is also prone to bacterial
contamination. Koji must be made using steamed rice
and steaming equipment is therefore essential for
making koji. You can use either a wooden or stainless-
steel steamer. Although a Japanese wooden steamer
(right) or Chinese bamboo steamer (left) is preferable,
a stainless steel steamer (back) can also be used. If
you cannot get hold of a steamer, you can use a deep
pan with a sieve placed inside. Pour water into the
pan, place the rice inside the sieve and boil the water.
Make sure the bottom of the sieve is not touching the
water. Alternatively, you can make a steamer yourself.
You will need a pan, a sieve, a thick wooden plate (wider
than the pan) and a wooden frame. First, make a hole
in the centre of the wooden plate (diameter 3-4 cm).
Fill the pan with water and place the wooden plate on

⑨ステンレスバット、トレイ…2枚

　蒸した米に麴菌を培養するために、蒸した米を入れておく容器、バットやトレイが必要です。日本では伝統的に「麴蓋」と呼ばれ、杉の木でつくられ、麴菌が繁殖しやすいように工夫された麴つくり専用のトレイがあります。普通の家庭ではステンレスか琺瑯のバットでいいでしょう。そのまま使ってもいいし、もし結露して蒸し米が濡れるなら、さらし布を敷くといいでしょう。今回の300gの米には幅22cm×奥行15.5cm×高さ3.5cmのステンレスバットを使っています。

⑩電気毛布（保温道具）

　麴菌を育てるには、30～40℃の温度と、70～90％の湿度を維持することが必要です。この環境を用意できれば、世界中どこでも麴はつくれます。温度のキープには、

top. Place the wooden frame on top of the plate and the sieve inside the frame. Bring the water to the boil, wrap the rice in a loose-woven cloth and put it inside the sieve. (Make sure the rice is not in direct contact with the wooden plate.) Place a lid on top of the frame to trap the steam inside.

9. Two stainless steel trays

One tray is used to hold the steamed and inoculated rice and the other (upside down) is used as a lid. Traditional trays are made of Japanese cedar, which has special properties that make it ideal for growing koji. However, stainless steel or enamel trays are fine for home use. If there is condensation, place a thin cotton cloth at the bottom of the tray to absorb excess water. In this book, I used a W22cm × D15.5cm × H3.5cm stainless steel tray for 300g of rice.

TIMINGS
タイムスケジュール

◎時間はあくまでも目安。例えば、蒸しあがりは「ひねりもち」（p25）をつくり、もちになっていれば米の中まで火が通っている合図。浸漬時間・蒸し時間は、季節、米の品種、精米具合、道具などでまったく変わります。時間やデータだけでなく、自分の五感を使い、実際の原料や麴の状態を見て判断しましょう。
◎麴菌は50℃以上だと死んでしまいますが、28℃以下だと胞子が発芽しません。温度計をチェックしながら作業は手早くします。

*Times indicated here are for guidance only. For example, when steaming you should always check the texture of the rice by hand ("hinerimochi" (p25)) instead of relying on the length of time indicated. The time required for soaking and steaming can vary depending on the season, rice variety, polishing rate or equipment used. So never rely solely on the times given, but instead train your five senses and use them as your guide.
*Koji will not survive over 50℃ (122°F) and will become dormant below 28℃ (82°F). Check the temperature constantly and work quickly.

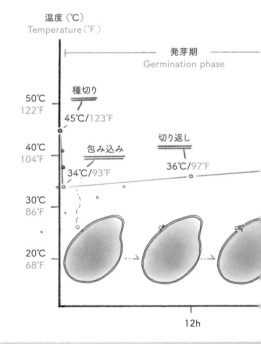

温度（℃）
Temperature（°F）

発芽期
Germination phase

50℃ 122°F — 種切り 45℃/123℃

40℃ 104°F — 包み込み 34℃/93℉ — 切り返し 36℃/97℉

30℃ 86°F

20℃ 68°F

12h

この本では電気毛布を使います。電気毛布のよいところ
は、世界中で買え、どんな形のものでも包めるので、量
が少ない場合も、多い場合も自由に調整できること、そし
てある程度、温度を調整できることです。使わないときに
は、小さく折り畳んで収納できるのでスペースを取りませ
ん。普通の人にとって麹つくりは、月に1回か、年に数回
でしょう。だから道具はできるだけ増やしたくありません。
最初はミニマルで、汎用性があるものでやるのがオスス
メです（湿度は蒸し米をバットで上下で挟んでおくことで、
「蒸し」で得た水分を逃がさず乾燥を防ぎ、麹菌にとっ
て必要な湿度を維持できます）。あれば、湯たんぽも使え
ます。

10. Warming equipment

When growing koji, you need to maintain the
temperature between 30-40°C (86-104°F) and
the humidity between 70-90%. Koji can be made
anywhere in the world as long as these requirements
are met. In this book, I used an electric blanket to
maintain the temperature. Electric blankets are
readily available in most countries and can be used
to wrap objects regardless of their shape. Some also
have built-in temperature control. I recommend using
electric blankets as they can be folded and will not
take up much space when stored. Most of you will not
be making koji every day, but perhaps once a month
or a few times a year. Taking up storage space with a
lot of koji-specific equipment is therefore not ideal
and I recommend using a minimum of general-purpose
equipment. You could also use hot water bottles.
I use two trays to maintain the correct humidity.
Placing them on top of each other to form a shell
keeps in moisture,
prevents the rice drying
out. In extremely dry
conditions, you may
want to place a damp,
tightly wrung cloth
over the tray as an
additional precaution.
Make sure the cloth is
not in direct contact
with the rice.

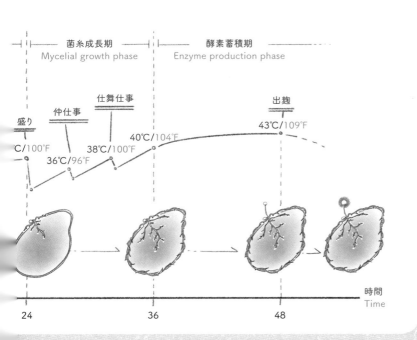

菌糸成長期
Mycelial growth phase

酵素蓄積期
Enzyme production phase

仕舞仕事

仲仕事

盛り

出麹

43℃/109℉

40℃/104℉

38℃/100℉

C/100℉

36℃/96℉

時間
Time

24　　　　　36　　　　　48

STEAMING RICE 米を蒸す

　麹をつくるには、まずは麹菌のすみかとなる蒸し米をつくります。麹つくりは蒸し米という培地に麹カビを生やす作業です。だから蒸し米の環境はとても大切。目指すは**外硬内軟**。これは、**表面はパラっとかたく、さわっても手につかず、内側には水分が充分にあって軟ら**かく保水した状態です。逆に、表面が軟らかく、さわると手にくっつく水分が多いベチャ飯は、雑菌が繁殖しやすく、失敗しやすいです。米は充分に吸水させ、水を切り、しっかりとした蒸気で蒸すのがポイントです。蒸し米の状態がその後の麹の質をほぼ決定します。麹つくりで一番大切な工程といえます。

　生米から蒸し米後の水分の増加量は34〜38％前後がオススメです。

The rice steamed in this step will be used to grow koji over the next three days. The quality of the steaming is therefore extremely important. Ideally, the rice should be **soft on the inside and firm on the outside**; this texture is called gaiko-nainan. In other words, you should aim to **steam the rice until it does not stick to your hands, but is still moist inside**. When the surface is soft and the rice sticks to your hands, it is prone to bacterial contamination and the process will not be successful. For proper steaming, it is important to soak and drain the rice well before steaming at high pressure. It is no exaggeration to say that the final product is determined by steaming, so this is the most important step when making koji. In general, properly steamed rice is 34–38% heavier than before cooking.

STEP. 1 Senmai

WASHING RICE

〜米を洗う

米は流水で、水が透明になるまでざっと洗う。ぬか分は麹菌の栄養にもなるので、「研ぐ」必要はない。

Wash the rice under running water until the water runs clear. You do not need to rinse the rice too much, as bran feeds the koji.

浸漬
<ruby>しんせき</ruby>

Shinseki

SOAKING RICE IN WATER

～水に浸けて吸水させる

12-24 HOURS

たっぷりの水に米を12〜24時間浸けてしっかり吸水させる。気温が高く水が温まったときは水を入れ替える。

Soak the rice in plenty of water for 12-24 hours until it has absorbed water all the way to its core. You may need to change the water if it warms up on a hot day.

Kashi

かし

GENTLY PRESSING AND ROLLING THE RICE BETWEEN TWO FINGERS TO SEE IF IT CAN BE CRUSHED TO POWDER.

～指で米を摺りつぶし、粉になるかをチェックする

米の吸水は麹つくりにおいて最も大切な工程。どれくらい吸水させるか、その時間は、米の品種、銘柄、精米具合、季節、地域、収穫年によって全く違う。同じ品種でも収穫年が違えば、吸水時間は変わるので一律に何分と時間では表せない。時間は参考にとどめ、常に直接米に触れて判断する。そのとき目安になるのが「かし」といって、吸水した米を指で摺りつぶしてみること。米を1粒指に取り、親指と人さし指で摺りつぶし、**簡単に米粉になるよう**なら中心まで充分に吸水している。摺りつぶしてみて、**硬く透明な塊が残る**ようならまだ中心まで吸水していない。

Soaking is another important step when making koji. The length of time required will vary depending on the rice variety, degree of polishing, season, region and year. This means soaking time can't be predicted, so you have to rely on your sense of touch to check. This action of checking is called "*kashi*", where you gently press and roll the rice using two fingers. Pick up a grain of rice and use the tip of your thumb to grind it against the second joint of your index finger. If the rice is completely soaked, **it should crumble to powder easily**. If there is still **a hard, transparent section in the centre**, it needs to soak for longer.

3

ザル上げ

Zaruage

DRAINING THE WATER

〜水を切る

ザルに上げ、5〜15分おいて水をしっかり切る。この時間は家庭用サイズの場合。米の量が増えれば時間は長くなる。その際、ザルは水平におくと下部に水がたまるので、斜め45度にしておく。金属製の蒸し器の場合、水切りが不充分だと蒸した米が軟らかく、表面がべたつくことがある。しかし乾燥した環境の場合、逆に水切り時間が長すぎても、表層の米が乾燥して硬くなり、蒸し時間が長くなる。**米の表面の水気はしっかり切る。かつ、乾燥はさせないことが大切。**

If using less than 1kg rice, drain it and leave it to dry in a sieve for 5–15 minutes. If using over 1kg, let the rice drain for longer. Tilt the sieve at a 45 angle, to avoid water collecting at the bottom. The rice may become too soft and sticky if it is not drained properly, particularly when using a stainless steel steamer. On the other hand, if you leave the rice to drain for too long (especially in dry conditions) the surface may become too dry and take longer to steam. The golden rule is to **drain excess water while preventing the surface becoming too dry.**

Put
the drained rice
in the steamer

〜ザル上げした米を蒸し器に入れる

　四つ折りにしたさらしを入れ（写真1）、その上に蒸し布をのせて（写真2）、水切りした米を入れる（写真3）。木製の蒸し器は米がぬれないので、下に敷くさらし布は必要ないことが多い。

Fold the light cotton/linen cloth in half twice and place it at the bottom of the steamer (picture 1). Then, place the loose-woven cotton/linen cloth on top (picture 2) and scoop the drained rice into the sieve (picture 3). You don't need the light cloth with a wooden steamer, as the wood absorbs excess water.

STEP. **4** 蒸し

Mushi

STEAMING RICE

〜米を蒸す

鍋に湯をたっぷり沸かし、蒸し器をのせて蒸し布を敷く（p21参照）。蒸し器への米の入れ方は、米の量や使う道具で若干違ってくる。目的は外硬内軟に蒸しあげることなので、結果を見ながらいろいろ試してみることが必要。もし米の量が1kg以下で少ないなら、一度に全部蒸し器に入れて蒸しても大丈夫。米の量が2kg以上とある程度多いなら「**抜けがけ法**」で蒸したほうが均一に蒸しあがる。

Boil plenty of water in a deep pan and place the steamer on top. Lay the loose-woven cotton/linen cloth in the steamer (refer to p.21) and put the rice inside the cloth. The best way of loading the rice in the steamer will differ depending on quantity and equipment used. Always keep in mind that your aim is to achieve the soft inside and firm outside required for making koji (*gaiko-nainan*) and adjust accordingly. For less than 1kg of rice, you can place it all inside the steamer at once. If steaming more than 2kg, use the layering method (*nukegake*, next page) to ensure even steaming.

NUKEGAKE (LAYERING)

抜けがけ法

一度に大量の米を入れて蒸すと、米の水分が下層にたまり、上層は熱が入らず蒸しムラができる。下層が水分の多い「ベチャ飯」になると、雑菌が繁殖しやすくなるので、何回かに分けて米を入れるのが「抜けがけ法」。

まず、蒸気の上がった蒸し器に1/3量の米をすき間なく敷き詰める。少し待つと米の間から蒸気が上がり、蒸気の抜けたところの米は半透明になってくる。そうしたら新たに残り1/3量の米をかける（写真）。再び蒸気が抜けたら同様に残りの米をかける。このように少しずつ蒸し器に米を入れていく。蒸気が米から抜けるのを確認しながら、米を「かける」ので「抜けがけ」という。今は専用の機械を使うことも多いが、プロの酒造りの現場でも蒸し器を使う場合はこのように蒸す。

Large quantities of rice put in the steamer at once may steam unevenly as steam becomes trapped at the bottom of the steamer and does not pass through to the top, leaving the rice at the bottom stickier. As mentioned above, sticky rice is vulnerable to bacterial contamination and so should be avoided. This problem can be solved by using layering method.

This method allows rice to steam evenly by loading it in several layers. Wait until steam comes through the steaming hole and load 1/3 of the rice in the steamer. Make sure there are no gaps at the edges or corners. Wait until the stream reaches the top (look for a change in colour, the rice will become translucent when the steam comes through). Repeat this step three times until all the rice is steaming in three layers. The name "*nukegake*" comes from the Japanese "*nukeru*" (to pass through) and "*kakeru*" (to pour). This reflects the sequence of actions where the rice

is poured into the steamer and you wait for the steam to come through the layer. Most sake breweries use automated steamer systems, but some still use this method with large commercial steamers.

米がすべて入り、表面から蒸気が抜けるのを待って、蒸し布をかぶせて、蒸し器の蓋をする。中火の強いまま、ここから40〜60分蒸す。ただし、蒸しあがりの時間は米の状態や蒸し器によって全く違うので、時間は目安として、必ず「ひねりもち」（p25）で確認すること。

When all the rice is loaded in the steamer, wait until the steam reaches the top of the final layer. Fold over the edges of the loose-woven steaming cloth to wrap the rice and close the lid. Steam the rice for 40-60 minutes over medium heat. Steaming time will vary depending on the condition of the rice, type of steamer used and other factors. Never rely exclusively on the timings given, always check the texture of rice by hand ("*hinerimochi*") (see p.25).

5

蒸しあがり

Mushiagari

CHECKING THE TEXTURE OF THE RICE

〜米を充分に蒸せたかチェック

米を充分に蒸せたかどうか、蒸しあがって
いるかどうか、「ひねりもち」をつくって確認する。

Check by hand if the rice is properly steamed.

Hinerimochi

ひねりもち

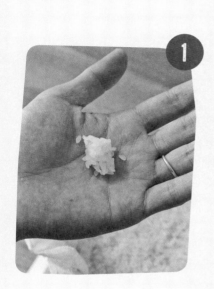

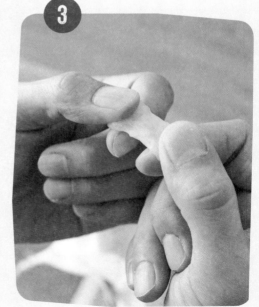

　熱いうちに米を少量手に取る（写真1）。両手のひらでひねりつぶす（写真2）。一粒ずつが溶け合いまとまる、引っぱって粘りのあるもちの状態になっていれば（写真3）蒸しあがり。まとまらない、表層が硬い場合はさらに10分蒸す。

While the rice is still hot, take a spoonful in your hands (Picture 1). Knead it with the base of your thumb (Picture 2). If the grains merge together and form a stretchy ball, the rice is properly steamed. If the grains are still crumbly or the surface is hard, steam for a further 10minutes.

INOCULATE WITH KOJI SPORES

ONE SPORE PER GRAIN

種麹をつける
〜米１粒に胞子１粒

　ここでは、適切に蒸された（外硬内軟に蒸しあがった）米を、麹菌が繁殖できる温度に冷まし、種麹（麹菌の胞子）をつけます。ここで種麹がつかなかった米は麹にはならないので**米1粒に胞子1粒のせてあげるつもりで丁寧に振り**かけ、よく混ぜます。ただし、米の温度が高すぎても低すぎても、また乾燥しすぎても麹菌は繁殖できません。暖かく風のない部屋で、手早く作業するのがオススメです。

After confirming the rice is properly steamed (soft on the inside and firm on the outside), wait until it is cool enough to sprinkle the spores. Any rice not inoculated with koji spores at this point will not become koji. Mix the rice well and aim to **sprinkle one spore onto each grain of rice**. Koji will only grow at optimum temperature, not if the temperature is too high or too low, or if the rice is too dry. This step therefore needs to be carried out swiftly in a warm room without air currents.

STEP. **6**　蒸し取り

Mushitori

REMOVING RICE FROM THE STEAMER AND COOLING IT

〜蒸した米を取り出し、粗熱をとる

　蒸し米をひねりもちでチェックしたら、清潔なスケッパーやスプーンと滅菌したバットを用意して、蒸し米を取り出し広げる。蒸し米はよく混ぜ、広げたりまとめたりしながら乾燥しないように気を配る。

　バットに広げた蒸し米は、外と内の米を入れ替えるように混ぜながら45℃以下まで冷ます。ここから**8**（包み込み）の作業までは30℃以上で終われるように手早くする。

After checking the texture of rice by hand, use a clean dough scraper or spoon to spread it on a sterilised tray. Separate the rice into individual grains by repeatedly spreading it out and gathering it up, without letting it get too dry.
Move rice from the inside of the tray to the outer edge until the temperature drops below 45°C (123°F). Complete the steps below up to number 8 (*tsutsumikomi*) before the temperature of the rice falls below 30°C (86°F).

バットを滅菌する

　使う直前にステンレスバットを滅菌する。バットを100℃のオーブンに入れて2〜3分焼くか、または熱湯をまんべんなくかけて清潔な布巾で拭いて乾かし、素手でさわれる、やけどしない程度に冷まして温かいまま使う。バットを温めてから使うと、熱い蒸し米を入れても、バットが結露しない。

STERILISE TRAYS

Always sterilise trays before use, either by heating them in the oven (2–3 minutes at 100°C/212°F) or by pouring freshly boiled water over them and wiping with a clean cloth. Let the trays cool until they are cool enough to touch and put the rice on to the trays while they are still warm. This avoids condensation when the warm rice is placed in the trays.

乾燥しやすい環境では「埋け飯」に

　乾燥しやすい環境や地域、季節、または表層の硬い米の場合は、混ぜながら冷ますのではなく、蒸し米が熱いうちに2枚目のバットで蓋をして（少しだけずらして湯気を逃がす）、外気に触れないようにし、水分が飛ばない状態でしばらくおいてゆっくり冷ます。もし室温が低いなら、このバットごとさらに毛布や電気毛布で包み保温しながらゆっくり冷ます。これを埋け飯という。軟らかめの蒸し米にする効果や、蒸し取り直後の、蒸し米の水分が熱で飛んで乾燥してしまうのを防ぐ効果がある。低精白の米や麦、または乾燥した環境では有効なので試してみるとよい。

COOLING RICE IN DRY CONDITIONS (*IKEMESHI*)

This method is recommended in dry conditions or when the rice is particularly hard. Instead of physically cooling rice by airing as described above, use the second tray as a lid. Place it over the first tray, slightly shifted so the steam is not trapped, and leave until the rice has cooled. You may need to wrap the trays with a blanket (or electric

blanket) to cool them down gradually. This method is particularly effective when you want to make the surface of the grain softer (useful for lightly polished rice or barley) or retain moisture (in extremely dry conditions).

扇風機で冷ましてはダメ？

　風を送りながら冷ますと表面が乾燥してしまい、麹菌の成長に必要な蒸し米の水分が減りやすい。すると麹菌の発芽にムラが出たり、菌糸の成長が遅れたりする。ここでは米の中の水分を残しながら温度を下げたいので、エアコンなども止めて、部屋の空気が動かない状態で作業する。

CAN I USE A FAN TO COOL THE RICE?

I don't recommend using a fan. Air currents dry up the moisture koji needs to grow, which can result in staggered germination or late growth. You need to reduce temperature without losing moisture. If you have air conditioning, turn it off to stop air currents.

Tanekiri

INOCULATING THE RICE WITH KOJI SPORES

〜種麹を振る

蒸し米に種麹を振ることを種切りという。45℃以下に冷めた蒸し米に種麹を振りかける。種切りの目的は、蒸し米に種麹を付着させること。すべての蒸し米に付くようイメージしながら、種切りを3回に分けて行なう（米1粒に胞子1粒）。茶こしに種麹を入れ、蒸し米に対して5〜10cmほどの高さから茶こしをやさしく叩きながら、1/3量ほどを表面全体に振る。これを3回繰り返す。

You're now ready to sprinkle the koji spores onto the steamed rice (to inoculate the rice with koji). Make sure the rice is below 45℃ (113°F). Try to get at least one koji spore on each grain of rice. To inoculate thoroughly, carry out this step in three stages. Hold the tea strainer 5 –10cm above the rice, gently tap to sprinkle 1/3 of the spores, and mix the rice. Repeat three times.

まず、1回目。表面全体に振りかけたら、スケッパーで全体をよく混ぜる。新しい蒸し米を表面に持ってきたら（写真1、2）2回目、同様に種麹を振り（写真3）、全体をよく混ぜる。3回目も同様にする。3回目は種麹を残さないようすべて使い切る。もし途中で種麹がなくなったら、新しい種麹を足してもよいので3回かけて蒸し米全体に振りかけるようにする。種の量は多少増えても構わない。3回目の種切りの後、もう一度全体をよく混ぜる（写真4）。種切りが終わったら、蒸し米の表面をきれいに平らにならす（凸凹のままだと温度・湿度が逃げやすい）。

Sprinkle 1/3 of the spores evenly over the rice and mix well using the dough scraper. Turn over the rice (Picture 1, 2), sprinkle another 1/3 of the spores (Picture 3) and mix well. Turn over the rice again and sprinkle the rest of the spores. If you run out of spores, add more from the packet. Using a small quantity of additional spores will not make much difference. Sprinkle and mix three times (Picture 4). Make sure the surface of the rice is even to avoid heat loss and evaporation.

包み込み

Tsutsumikomi

WRAP FOR INSULATION AND TO ENCOURAGE GERMINATION

〜包んで保温し、発芽を促す

発芽期
GERMINATION PHASE

温度計のセンサーの先端を、蒸し米の真ん中に差し込み（写真1）、そこにさらにもう一つバットをかぶせて蓋にする（写真2）。2枚のバットごと電気毛布で包んで（写真3）、蒸し米の温度が34〜38℃で12〜18時間保温する。最初は電気毛布の温度は中か強。その後、温度を見ながら調整する。湯たんぽを使う場合は、バットごと毛布で包んで上下に湯たんぽ2個をおいて挟むとよい。

Insert the thermometer into the centre of the rice (Picture 1) and place the second tray upside down as a lid (Picture 2). Wrap the trays with an electric blanket (Picture 3). Keep the temperature at 34-38°C (93-100°F) for 12-18hours. Set the heat to medium to high at first, then adjust according to the measured temperature. If using hot water bottles, use two. Wrap the trays with a blanket and place one water bottle above and another beneath.

Kirikaeshi 切り返し

TURNING TO KEEP TEMPERATURE AND HUMIDITY EVEN

～温度・湿度を均一に保つ

　包み込みから12～18時間で、一度「切り返し」をする。切り返しは麹菌の発芽の途中で混ぜて、蒸し米全体の温度・湿度・酸素の条件を均一にすることが目的。麹菌の成長するスピードは温度・湿度・酸素によって変わるので、環境条件にムラがあると麹のできあがりもムラができる。そして発芽のタイミングがずれると、最後までそのギャップを取り戻すことは難しくなる。

　蒸し米の入ったバットを電気毛布から出し、蓋（バット）を開け、温度計をはずし、香りや味を確かめる。栗のような香り、穀物の優しい味がすれば麹菌の成長は順調。蒸し米の温度と湿度が均一になるように手早く全体を混ぜる（写真1）。バット全面に広げて表面を平らに整える（写真2）。酸っぱい香りがしたり、はっきり甘みを感じるときは水分が多すぎるので、乾燥させる。アルコール臭や揮発したような香りがするなら、手を入れてよく酸素を補給し、通気性をよくする。「切り返し」をしたら、温度計を差し、再びバットをかぶせて、電気毛布に包み、保温を続ける。

　この後、種切りから24～26時間後の作業「盛り」まで34～38℃前後の温度帯から36～40℃まで徐々に温度を上げる。

Turn the rice once 12–18hours after wrapping. This ensures even temperature, moisture and oxygen supply during the germination phase, all factors that influence speed of growth and the quality of the final koji. If the spores germinate at different times, this creates differences that will impact the whole process.

Take out the tray and take off the lid. Remove the thermometer and check the aroma and taste. If the koji is growing properly, it should smell like chestnut and have a subtle hint of sweetness. Swiftly mix the rice to even out the temperature and humidity (Picture 1). Spread out the rice in the tray and make the surface even (Picture 2). If the rice smells sour or is very sweet there is too much moisture and it needs to be dried out. If the rice smells like alcohol or volatile compounds, it needs more oxygen. Mix well to supply oxygen. Put the thermometer and lid back in place, wrap the trays with the electric blanket and carry on keeping them warm.

Gradually increase the temperature to 36–40°C (97–104°F) from now until the first mixing stage, usually 24–26hours after inoculation.

温度が上がりすぎたら？

　途中で45℃以上になりそうなら、電気毛布から出して全体をほぐし混ぜ、冷気に当てる。これで米の温度も下がり、全体の温度と湿度が均一になる。再び電気毛布で包んで保温する。麹菌の生育温度は28～45℃。冷え過ぎても死ぬわけではないので、温めればリカバリーできる。ただし50℃以上になると麹菌が弱るか死んでしまう可能性があり、リカバリーは難しい。温度は上がりすぎないよう、途中で何度か温度計をチェックする。とくに初めての人は、1～2時間おきに見るとよい。

WHAT IF THE TEMPERATURE GETS TOO HIGH?

When the temperature approaches 45°C (113°F), take the trays out. Mix the rice to supply cool air. This should reduce the overall temperature and even out the humidity. Wrap the trays with the electric blanket again and keep warm. The optimum temperature for koji is 28–45°C (82–113°F): it doesn't die at low temperatures and becomes active again when inside its temperature range. However, if the temperature goes over 50°C (122°F), the koji may become less active or even die. Check the thermometer occasionally to avoid the temperature getting too high. If it's your first time making koji, check the temperature every few hours.

MYCELIAL GROWTH 菌糸を成長させる

ここからは菌糸をのばす麹菌の成長期です。麹菌が育っていく過程で発熱するのですが、それが一気に進むと、発酵熱で菌が死んでしまうので、途中手入れをして、全体の温度と湿度を均一にします。蒸し米全体の環境を均一にすることで、米1粒1粒に対して、表面にも内側にも菌糸がしっかり伸びてよい麹ができます。

This phase is when koji grows mycelia. Growing mycelia generates heat, and if the koji grows too fast it could be killed by the heat it generates. To avoid this, mix the koji to limit its growth and even out overall temperature and humidity. High quality koji has strong mycelial growth both inside and outside the rice grain, which requires even temperature and humidity.

STEP. 9 盛り（もり）

菌糸成長期
MYCELIAL GROWTH PHASE

Mori

FIRST MIXING

〜 1 回目の手入れ

培養24時間後〜、温度36℃以上。菌糸の成長が見た目で表面1〜2割、麹っぽい香りが立ってきたら、「盛り」に入る。蓋（バット）を開け、温度計をはずし、米全体をまんべんなく混ぜる。混ぜることで酸素を供給し、温度と湿度の条件を均一にする。

2日目は「盛り・仲・仕舞」といって温度を確認しながら3回手を入れて、混ぜる作業がある。この作業を地方によっては「1番手入れ・2番手入れ・3番手入れ」ともいう。

「盛り」は、伝統的な方法である「布で包み込んで発芽させた麹を、木の麹蓋に移し、容器の真ん中に山のように盛りあげる作業」からきた名前。バットの場合は、麹は移さずに、最後まで同じバットでそのまま培養を続ける。

Perform this step around 24hours after inoculation, when the temperature has risen to 36°C (97°F). At this point, 10-20% of the surface of the rice should be covered with mycelia and the koji should start emitting a distinctive, subtly sweet smell. Open the tray and remove the thermometer. Mix the rice well to even out the temperature, humidity and oxygen supply. Koji is mixed three times while being grown (typically on the second day). The mixing steps are called *mori*, *naka* and *shimai*. The timing of each mixing should be guided by the temperature of the rice.
"*Mori*" means "heap" in Japanese. The name comes from the step in traditional koji making where germinated koji is unwrapped and transferred in heaps to wooden trays.
The method introduced here is a simplified one, so the koji stays in the tray throughout the process.

麹が湿っぽい・発育が悪い場合

米の量と環境によって、麹が湿っぽかったり濡れたりして、部分的に麹の発育が悪かったりする場合がある。湿気を取るためにバットと麹の間に木綿の布を1枚はさんで余分な水分を取り、蒸し米が濡れないようにするとよい。

WHAT IF THE KOJI IS TOO WET OR ISN'T GROWING WELL?

The koji may become too moist or even wet and you may find the growth is uneven. If this happens, lay a light cloth underneath the koji to absorb excess moisture.

麹が乾燥する・硬くなった場合

麹つくりの途中で、蒸し米が乾燥で硬くなり、生育が悪い場合は、直接、蒸し米に触れないように、バットに同じ大きさのザルをかぶせてから（写真1）水に濡らして硬く絞った布をかぶせ（写真2）、さらに蓋をする（写真3）。

麹を見ながら、麹にとって最適な環境に常に調整し続けることで、麹つくりの技術も、麹の質も向上していく。

WHAT IF THE KOJI IS TOO DRY OR BECOMES HARD?

The koji may become too dry and hard and you may find this reduces growth. If this happens, place the sieve (roughly the same size as the tray) upside down (Picture 1) and lay a damp, tightly wrung cloth on top (Picture 2) then put the lid back on (Picture 3). The sieve keeps the rice out of direct contact with the moist cloth.
Constantly observing the koji and maintaining optimum conditions will improve both your koji making skills and the quality of the final koji.

STEP. 10

仲仕事 (なか)

Naka

SECOND MIXING

〜2回目の手入れ

　培養28〜30時間、盛りから4〜6時間後、温度が36〜38℃になったら、2回目の手入れ「仲仕事」を行なう。菌糸は表面3〜5割に成長。作業は盛りと同じ。全体をまんべんなく混ぜて、酸素を供給し、温度と湿度の条件を均一にする。仲仕事後はまた蓋をして保温し、培養を続ける。

Perform this step around 28–30hours after inoculation (4–6hours after first mixing), when the temperature rises to 36–38℃ (97–100°F). At this point, 30–50% of the surface of the rice should be covered with mycelia. The process is the same as the first mixing, mix the koji well to even out the temperature, humidity and oxygen supply. Put the lid back on and keep warm.

11

Shimai

THIRD MIXING

～3回目の手入れ

　培養32〜34時間、仲仕事から4〜6時間後、温度が38〜40℃になったら3回目の手入れ「仕舞仕事」を行なう。菌糸は表面6〜8割に成長。だんだんと麹っぽい発酵香がしてくる。麹は栗のような香りがする。全体をまんべんなく混ぜて、酸素を供給し、温度と湿度の条件を均一にする。仕舞仕事後は容器に均一に麹を広げ、2本線で溝を引いて畝をつくる。これを「花道」と呼ぶ。花道は、表面積を広げ、熱と水分を発散させ、酸素を供給し、後半の麹の成長を助ける役割がある。その後、蓋をして保温し、必要ならp33のような湿らせた布をかけて保湿し、培養を続ける。

　これ以降はもう麹は混ぜない。ここからは菌糸が外側へも活発に成長するので、さわりすぎると、菌糸の成長を邪魔する。

Perform this step around 32–34hours after inoculation (4–6hours after second mixing), when the temperature rises to 38–40°C (100–104°F). At this point, 60–80% of the surface of the rice should be covered with mycelia and the distinctive sweet, chestnut-like smell of koji should be more apparent. Mix the koji well to even out the temperature, humidity and oxygen supply. Spread the rice evenly in the tray and use your fingers to make two troughs. The troughs are called *hanamichi* (flower road). The troughs enhance koji growth by increasing surface area, helping to disperse heat and moisture, and supplying oxygen efficiently. Put the lid back on and keep warm. Use a moist cloth (refer to p.33) as necessary.
This is the final mixing. After this point the mycelia are actively growing and mixing too much would hinder them.

最高温度をチェック

Checking
maximum temperature

酵素蓄積期
ENZYME
PRODUCTION
PHASE

培養36～38時間後、仕舞仕事から4～6時間、温度が38～43℃になったら、麹の中に酵素が多く生産され、蓄積される時期になる。この時間と温度を確認したら、このままの温度を8～12時間キープ。ここではもう混ぜない。温度と湿度をキープするだけ。ただ乾燥には気をつける。

このとき40～43℃で12時間維持すると、グルコ・アミラーゼ（でんぷん分解酵素）が多く生産され、甘みを強くつくる麹になる。特に43℃を厳密に維持するとピュアな甘さになる。38～40℃ではアルファ・アミラーゼ（でんぷん分解酵素）が多く生産され、コクのある甘さになる。30～36℃を維持すると、プロテアーゼ（たんぱく質分解酵素）が多く生産され、旨みを多くつくる麹になる。38℃±2前後を維持するとアミラーゼとプロテアーゼの両方がつくられる。ここの温度と時間をコントロールすることで、酵素のバランスを調整できる。このバランスが最終的な発酵食品の味をつくる。

米の状態の変化

生の米から完成までの米粒で状態の変化を見てみる。だんだん白く濁ってきて、表面に菌糸が生えてくるのがわかる。

At around 36-38hours after inoculation (4-6hours after third mixing), when the temperature rises to 38-43℃ (100-109°F), koji starts producing and accumulating enzymes. Once it reaches this stage, keep the temperature stable for another 8-12hours without mixing. Maintain the temperature and humidity while making sure the koji does not dry out too much. If you maintain the temperature at 40-43℃ (104-109°F) for 12hours, the main enzyme will be glucoamylase, which breaks down starch. The resulting koji will be suitable for uses that require sweetness. If the temperature is maintained at 43℃ (109°F) throughout, the final product will be pure sweetness. If the temperature is maintained at 38-40℃ (100-104°F), the main enzyme will be alpha-amylase, another enzyme that breaks down starch. This enzyme contributes a rich, robust sweetness to the final product. If the temperature is maintained at 30-36℃ (86-97°F), the main enzyme will be protease, an enzyme that breaks down protein. The resulting koji will be suitable for uses that require umami. If the temperature is maintained at 38℃ (100°F)±2, both amylase and protease are produced. This is how you can adjust the balance of enzymes, which in turn determines the taste of the final fermented products.

CHANGES IN APPEARANCE DURING KOJI GROWTH

The following pictures show how the appearance of the rice changes from uncooked rice to rice koji. As mycelia grow inside and outside of the grains, the rice turns whiter and fuzzier.

生米
UNCOOKED RICE

蒸し米
STEAMED RICE

切り返し
TURNING
(KIRIKAESHI)

盛り
FIRST MIXING
(MORI)

仲仕事
SECOND MIXING
(NAKA)

仕舞仕事
THIRD MIXING
(SHIMAI)

完成
COMPLETED RICE KOJI

COMPLETED RICE KOJI できあがり

米の中まで菌糸が入り込み、米は白く濁り、表面に菌糸が伸び、麹が完成です。ただ、できた麹はそのまま放置しておくと麹菌が成長し続け、自家発熱によって温度が上がって胞子がつくことがあります。容器から出して冷まして乾燥させ、成長を止めることで、麹として使えます。さらに乾燥させれば保存ができます。

Koji growth is complete when the mycelia have grown abundantly both inside and around the grain. At this point, the entire rice grain should be fuzzy and covered with matt white mycelia. If you carry on maintaining the heat and humidity, the koji will keep growing and generating heat, the temperature will continue to rise and eventually spores will be produced. To stop growth, take the koji out of the tray to cool and dry it. Properly dried koji keeps for longer.

STEP. 13 Completed Rice koji 完成

蓋を取って確認する。表面が菌糸でモフモフになり、さわるとふわっと盛りあがっていて弾力のある状態。

Remove the lid and observe the koji. The surface should be covered with fluffy mycelia and should feel soft, spongy and resilient.

完成した麹。米は白濁して表面に白い毛が生えている。これが麹の菌糸

Fully grown rice koji. Matt white surface with fine hairs, which are the koji mycelia.

STEP. 14

出麹 <small>(でこうじ)</small>

Dekoji

REMOVING KOJI FROM THE TRAY

〜容器から取り出す

　培養48〜60時間、酵素蓄積期の開始から12〜16時間後、菌糸が充分（表面9〜10割）に成長し、米を割ってみると内側にもしっかり菌糸が食い込んでいる。香りと味は栗のような「栗香（くりか）」「栗味」。こうなればできあがり。保温を解いて蓋（バット）と温度計をはずし、麹を取り出す。保温と保湿を止め、麹の成長を止めることを「出麹」という。培養時間が指定より短くても、胞子ができて色がついてきたらもう出麹したほうがよい（醤油用ならあえて胞子形成まで成長させることもあるが、甘酒や酒、味噌をつくる場合は、胞子に色がつくとできあがりの甘酒や酒に色がついて使いづらくなるし、雑味や苦味の元になるのであまり成長させないほうがよい）。

At around 48-60hours after inoculation (12-16hours from the beginning of the enzyme production phase), the rice should be almost fully covered with mycelia (90-100%). The mycelia should also have grown abundantly into the rice (check by breaking open a grain). The rice should smell and taste like chestnut. When you see this, it means that koji growth is complete. Stop warming, remove the lid and thermometer and take the koji out of the tray. You can take it out earlier if you find spores are being produced and the koji is changing colour. Although spores are intentionally produced in some cases (such as when making soy sauce), they are generally undesirable as they affect the colour of the final product and create impurities or bitter flavours (especially undesirable for amazake, sake and miso).

くずした麹を両手で包
んで鼻を近づける

Hold the separated koji with
both hands and smell

ここからすぐに味噌や甘酒に使えるし、乾燥させれば保存もできる。保存する場合は「枯らし」に入る。出麹は麹のできあがりを確認して、培養を終わらせるタイミング。これも実際の麹の様子を見て判断する。培養の標準時間は48時間ほどだが、43〜60時間の幅がある。温度が下がったり乾燥したりすると成長は遅くなり、温度や湿度も高いと成長は早くなる。一般的な白米でつくる白い麹の場合約60時間（それでも胞子が増殖し始めたら止めたほうがよい）までは培養できる。

You can use the koji immediately for miso and amazake. Drying it will preserve it for longer. Observe the koji to see when to stop its growth. The average time for making koji is 48 hours, but it can range from 43-60 hours. Growth takes more time if the temperature or humidity is low, or less time if the temperature or humidity is high. Koji can be grown on white rice for up to around 60 hours (but stop growth if you see spores appearing).

両手をすりあわせて
麹をバラバラにする

Separate the koji by
rubbing with both hands

枯らし

Karashi

DRYING THE KOJI

〜乾燥させる

できあがった麹をバットから取り出し、清潔な布やザル、乾いたバットに広げ、温度と湿度を飛ばして乾燥させることを「枯らし」という。麹の成長は水分がなくなり、乾燥することで止まる。布に広げて2〜3日風通しのいい涼しいところにおく。日本の夏や梅雨時期は、常温に放っておくと麹菌の成長がすすみ胞子がつくられ色がつくことがあるので、枯らしの最中も何度か混ぜるか、湿気の多いところにおかないよう注意する。

Drying involves removing koji from the tray, spreading it onto a clean cloth, sieve or dry tray to cool it down and drying it. Koji stops growing without moisture. Leave the koji in a cool, well-ventilated place for a few days. In humid conditions, such as Japanese rainy season or summer, you may find koji starts growing again and producing spores (and therefore changes colour) at room temperature. You can prevent this by mixing it regularly or moving it to a drier place.

Store

麹が手につかないくらいにパラパラに乾燥
したら密閉袋に入れて、保存できる。常温で
2週間、冷蔵庫で6ヵ月、冷凍庫で1年ほど
持つ。野菜用の乾燥機などで完全に乾燥さ
せて、密閉すれば常温でも1年以上持つ。

Dried koji can be stored for two weeks at room temperature,
6 months in the fridge and 1 year in a sealed bag in the
freezer (it should be dried until it does not stick to your
hands). Koji dried in a food dehydrator and packed in a
sealed bag can last for over a year at room temperature.

麹菌はカビ

　では麹つくりを一通りやったところで、生き物としての麹菌の生態を見てみましょう。麹菌の生態を知ることは、麹つくりの原理を知ることです。原理を知れば、どのような環境や、原料、道具でも原理を応用して麹をつくることができます。

　まず、麹菌はカビです。あなたがよく見る食べものに生えるカビと同じ生態です。有機質に生え、風呂場や水場のような、風の通らない暖かくて湿った場所が好きです。なぜ麹菌は湿った場所が好きなのでしょう。それは酵素が関係しています。麹の酵素は「加水分解酵素」。水分のある環境でモノを分解する酵素です。麹が酵素を生産するのは、麹にとっての食べものを得るため。つまりエネルギーを摂取するためです。

酵素で分解して栄養を摂る

　麹菌は栄養を摂取するために、根っこに当たる菌糸の先端から酵素を出します。菌糸を伸ばし、でんぷんやたんぱく質に触れると、そこでその対象物に対応した酵素が生産され、でんぷんやたんぱく質が分解され、糖やアミノ酸となるわけです。

　麹菌の酵素は「加水分解酵素」なのでこのとき、環境に水分が必要です。水分がなければ酵素は働かないので、対象物が分解されません。すると、麹菌は食べものが得られない→栄養が摂れない→だから成長できない、となります。充分な水分がある、つまり湿った

KOJI IS A MOULD

Now that you understand how to make koji, I want to take a closer look at koji's biology. If you understand koji biology, you start to understand the fundamental principles of making koji. Knowing those principles gives you the flexibility to make koji in any environment, with any materials or equipment you have to hand.

Koji is a family of microorganisms classified as moulds, which are often found growing on food. Koji feeds on organic matter and prefers warm, humid and poorly ventilated habitats. Why does koji mould like humid environments? This is related to the enzymes mentioned earlier. The enzymes produced by koji are categorised as "hydrolases", which break down their targets in the presence of water. Koji produces these enzymes in order to survive, or in other words, to obtain energy.

USING ENZYMES TO DIGEST FOOD

Koji produces enzymes at the tip of its mycelia to digest its food. When the mycelia sense starch or proteins, they produce corresponding enzymes to decompose them into sugars or amino acids.

As mentioned above, the enzymes produced by koji are hydrolases that require moisture to function. They are inactive in the absence of water, meaning koji cannot digest food or obtain nutrition and therefore cannot grow. On the other hand, when water is available, koji enzymes are active and start decomposing their substrates. Koji absorbs the decomposed

環境だと、その水分を利用して、酵素が作用し対象物が分解されます。小さな単位まで分解された栄養を、麹は菌糸の先端から吸収し、そこにまた菌糸を伸ばします。そして、でんぷんやたんぱく質にぶつかったらまた酵素を出し→分解し→吸収し→成長して→菌糸を伸ばし→また酵素を出す。これを繰り返しながら麹菌は成長していきます。麹つくりの最初の工程である浸漬で「米の中心までしっかりと水分を吸わせる」ことは、麹菌の酵素が働きやすい環境をつくることなのです。米が水を吸っていなければ、麹菌は菌糸を伸ばせません。麹つくりで湿度を維持するのは、菌糸をしっかり伸ばすためです。

外硬内軟の蒸し米で他の菌を抑える

そして蒸し米を外硬内軟にするのは、培地である蒸し米を他の菌に汚染させないためです。米が外側まで軟らかいと他のいろいろな菌も米を侵食しますが、外硬内軟の状態は、表面が乾燥して硬いので、他の菌が繁殖しづらいのです。でも麹菌には他の菌と違い菌糸があるので、水分のある米の内側に菌糸を伸ばせるので、他の菌を抑えて優位に成長することができます。ただ、麹菌の成長には酸素も必要で、軟らかく表面水分の多い蒸し米（ベチャ飯）は、米同士がくっつき酸素に触れる面積が少なくなるので麹菌は成長しづらくなります。また菌体が水に埋もれて酸欠になったりするので苦手です。外硬内軟にすることで、他の菌を抑えられ、麹菌にとって有利な環境が用意されているのです。

products (i.e. smaller molecules) from the tip of its mycelia, which in turn enables the mycelia to grow into newly formed cavities in the substrate. Koji grows by repeating these steps: encountering and sensing starch and proteins » secreting enzymes » decomposing substrate » absorbing small molecules » growing mycelia.

I explained earlier how thoroughly soaking the rice is the most important step when making koji. This is because it creates the optimum condition for koji enzymes. Koji cannot grow its mycelia into the rice unless the grain is properly soaked. I also repeatedly mentioned that maintaining the level of moisture is important. This is for the same reason: it creates the best environment for enzymes to work and mycelia to grow.

SOFT INSIDE AND FIRM OUTSIDE (*GAIKO-NAINAN*) TO PREVENT CONTAMINATION

Why do we aim to make the rice soft inside but firm outside when steaming? One reason is to prevent bacterial contamination. If the outer surface of the rice was soft and moist, other microbes would be attracted to it. However, they find it difficult to grow when the surface is dry and hard. But this isn't a problem for koji and its mycelia: they can penetrate the surface and reach inside to where there's plenty of moisture. Another reason why the surface should not be too soft and sticky is because koji needs oxygen to grow. When rice sticks together, surface area is reduced and oxygen is in short supply, making it hard for koji to grow. Wet surfaces can also submerge and suffocate the koji. This soft inside, firm outside state is the best one for growing koji while minimising the risk of contamination.

BASIC KOJI BIOLOGY

酵素の量は菌糸の成長具合で判断

　「麹つくりは酵素つくり」と最初にいいましたが、ここでわかるように麹菌の酵素は、菌糸の成長とともに生産されます。酵素は目で見えませんが、菌糸の成長具合を見ることで、どれくらい酵素が生産されたのかを予想することができます。麹のでき具合は、菌糸がどこにどれだけ成長しているのかを見て判断します。

　昔からいわれる麹のつくり方や、職人の仕事も、こうやって麹菌の生態から見るとすべてに理由があります。この「なぜそうなのか？」の原理を知って実際に作業をして理解すると、麹つくりはとても自由になるし、麹菌にとって心地よい環境を、人間が先回りして整えてあげられるので、麹菌はのびのびと成長することができます。「よい麹を育てることは、よい環境を整える」ということです。

EVALUATE ENZYME STRENGTH BY MYCELIAL GROWTH

In an earlier chapter, I explained that the purpose of making koji is to produce enzymes. Enzymes are produced when mycelia grow, and although they are not visible to the human eye you can estimate the quantity of enzymes produced from the volume of mycelia. Therefore, success when making koji can be evaluated by the volume and location (inside and outside the grain) of mycelia.

When you understand the biology of koji, all the details of the traditional koji making process start to make sense. With this understanding and practical knowledge from experience, you will be amazed by the flexibility of making koji. We can create a mutual relationship with koji: proactively providing optimum conditions lets the koji grow abundantly and produce as many enzymes as possible, which is beneficial for us.

CHAPTER 3

Using koji <u>in</u> <u>your everyday</u> <u>cooking</u>

Now you know how to make koji, let's start using it!
The recipes in this section are for fermented foods and condiments you can use in your everyday cooking. Some can be made and eaten immediately, while others take some time to ferment. Please try them and enjoy koji's fantastic benefits to the full.

麹でつくる

さて、麹ができたら
いよいよ発酵食づくりにとりかかりましょう。
ここで紹介するのは麹を使った
日常的につくれる発酵食と調味料です。
すぐできるもの、時間のかかるものがありますが、
日々の暮らしに取り入れて
発酵ライフを始めてみましょう。

甘酒

甘酒は、麹の酵素で米のでんぷんを糖化させてつくる、甘い穀物ドリンクです。酵素によって分解されたブドウ糖やアミノ酸、ビタミンなどが豊富に含まれ、人間にとって吸収しやすくなっています。温めても氷で冷やしても、柑橘やジンジャー、スパイスを入れても美味しいです。

Amazake is a grain-based, naturally sweetened drink made with koji's starch-degrading enzyme, amylase. It contains lots of glucose, amino acids and vitamins released by the enzymes in easily digestible forms. You can enjoy it warm, or cold with ice. I also recommend trying it with citrus fruit, ginger or spices for a refreshing drink.

Amazake

つくり方

麹のみでつくる方法と、ご飯と合わせる方法の二つがあります。麹のみなら、麹：水分＝1:1です。あっさりとした甘みと麹の香りが強くなります。ご飯と合わせるなら麹：ご飯：水分＝1:1:2です。甘みが強く、コクと旨みも出ます。

まず65℃の湯を用意します。そこに常温の麹、または麹とご飯を入れてよく混ぜ、55〜60℃で8〜12時間保温します。途中で1〜2回撹拌すると分解はより均一に早くなります。甘みがしっかり出れば完成です。瓶や保存容器に移し、冷めたら冷蔵庫に保存します。1ヵ月ほど持ちます。

How to make amazake

There are two different ways to make amazake. The first uses just koji, and the second uses koji and cooked rice. If you make it with just koji, mix koji to water in a 1:1 ratio. This amazake is subtly sweet and bursting with koji flavour. If you make it with koji and cooked rice, mix koji, rice and water in the ratio 1:1:2. This amazake is sweeter, full of flavour and has more umami.

To make amazake, first warm the water to 65℃ (149 °F). Then add the koji (at room temperature), or koji and cooked rice. Mix well and keep at 55-60℃ (131-140°F) for 8-12 hours. Mix several times to ensure even fermentation and speed up the process. When the amazake is sweet enough, take it away from the heat. Transfer to a container and cool, then it will keep in the fridge for about 1 month.

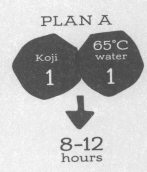

PLAN A

Koji 1 — 65℃ water 1

8-12 hours

PLAN B

Koji 1 — 65℃ water 2 — Rice 1

8-12 hours

乳酸発酵 Lacto-fermented

甘酒 amazake

手づくりの甘酒にはわずかながら乳酸菌が含まれていますが、乳酸発酵甘酒はその乳酸菌を培養した甘酸っぱい甘酒です。米でつくるヨーグルトのようなもので、さわやかな酸味と甘みがあります。市販の甘酒は殺菌されているので、これは手づくりの甘酒だからこそできます。乳酸と乳酸菌による整腸作用があり、麹由来の酵素やビタミンも含まれた栄養のある飲みものです。糖分と乳酸菌が組み合わさった甘酸っぱさが、根源的にヒトを魅了します。乳製品を食べない人への乳酸菌発酵食品としてもいいでしょう。

Homemade amazake contains bacteria that produce lactic acid, and this recipe uses them to produce lacto-fermented amazake. This "rice yoghurt" is full of refreshing sourness and sweetness. It can only be made using homemade amazake, as commercial amazake has been pasteurised and is no longer live. Lactic acid and the bacteria that produce it help to maintain gut health. This is a very nutritious drink as it contains plenty of koji-derived enzymes and vitamins. The combination of natural sugars and lactic acid creates a delightful taste, with a perfect balance of sweet and sour flavours – it will definitely stimulate your taste buds! It's a perfect alternative to dairy-based, lacto-fermented products.

つくり方

甘酒を瓶に入れます。このとき、瓶の口ギリギリいっぱいまで入れ、蓋をして瓶内の空気の量を少なくします。蓋はスクリュー式がいいでしょう。その後、それを冷蔵庫へ入れ7日ほどおきます。冷蔵庫から取り出し、いったん蓋を開けてみて様子を確認します。この時、甘酸っぱい乳酸香がしていればいい感じです。もし香りが感じられなくても、そのまま培養を続けます。

その後は常温で3～5日、変化が起きるまでおきます。直射日光の当たらない台所の隅などがいいでしょう。毎日1回、蓋を開けながら様子を確認します。見た目、香り、表面の色、そして味。暖かくなると急激に変化します。それが自分の感性にとって好ましいものなら順調でしょう。徐々に酸味が出てくると思うので、その酸味が自分にとって心地よいタイミングを見計らい、また冷蔵庫に保存します。あとは、自分の好きな時に飲むといいでしょう。保存は冷蔵庫で1ヵ月ほどです。

How to make lacto-fermented amazake

Fill a jar to the brim with amazake (leave as little air as possible) and close the lid. I recommend using a jar with a screw-on lid. Keep the jar in the fridge for about 7 days. Take it out and open the lid to see if you can smell a pleasant, sour-sweet lactic aroma. If you don't smell this aroma, close the lid and put the jar back in the fridge.

If you smell the right aroma, leave the jar at room temperature for 3-5 days. Keep it out of direct sunlight, for example in a corner of your worktop space. Check the appearance, aroma, colour and of course taste on a daily basis. These factors could change suddenly in warm conditions. Use your instincts to judge whether each factor is fully developed or needs more time – the sour flavours will gradually become stronger. When you are happy with the taste, store the jar in the fridge and enjoy the lacto-fermented amazake whenever it tickles your fancy. It will keep in the fridge for about 1 month.

Miso

味噌（みそ）

つくり方

How to make

麹をたっぷり使い、甘みと旨みを多めにつくる配合です。初めてでもつくりやすく、また食べやすい味噌で、味噌づくりの入門にちょうどよいでしょう。通常のレシピの2倍の麹を入れているのは、自分で麹をつくるからこそ。2回目以降は調整して自分の好きなバランスの配合を見つけてください。

This recipe uses a comparatively high ratio of koji (twice the amount of a standard recipe) and makes sweet, umami-rich miso. You made the koji yourself, so don't be shy about using it! This recipe is particularly suitable for anyone who is new to making miso, as it's easy to make and can be used in a wide range of dishes. After you've made it once, adjust the ratio of ingredients the next time to make a miso suited to your taste.

〈材料〉2.3～2.5kg
大豆…500g（煮て1000～1100g）
米麹…1000g
塩 …200～210g（煮た大豆＋麹の10％重量）
〈道具〉
大鍋、ザル、マッシャー、
3ℓ以上の蓋つきの容器

<Ingredients> To make 2.3-2.5kg of miso
Soybeans: 500 g (1,000-1,100g when boiled)
Rice koji: 1,000g
Salt: 200-210g
    (10% of total weight of boiled soybeans and koji)
<Equipment>
Large pan, sieve, masher, container with lid (3L or larger)

1 一晩浸水した大豆を2～3時間煮る。親指と薬指で簡単につぶれるほどの軟らかさになったら、ザルにあげる。煮汁は捨てずにとっておく。

2 豆を煮ている間に、麹をほぐして麹の10％の塩を混ぜておく。

1. Soak the soybeans overnight, then cook for 2-3 hours. When they are soft (easily squashed between your thumb and ring finger), drain them. Keep the cooking water as you may use it later.
2. While the soybeans are cooking, crumble the koji and mix it with some of the salt (10% of weight of koji).
3. Weigh the cooked soybeans and add more salt (10% of the weight of cooked soybeans). Squash the beans

3 1のゆであがりの大豆の重さを量り、10％の塩を加え、熱いうちにマッシャーなどでつぶす。

4 3が60℃以下になったら2を加えて混ぜる。水分が足りなければ少量の1の煮汁を加える。煮汁を加えたときは煮汁の10％の塩を混ぜる。

5 手で丸めて味噌玉をつくる。

6 味噌玉を投げつけるようにして容器に詰める。すき間ができないように最後に上から押して空気を抜く。

7 表面をならす。その際、熟成した自家製味噌があれば表面に塗る。自家製味噌には発酵に必要な菌がすでに入っているので塗って蓋をすることでよい発酵系の微生物を種継ぎでき、カビの発生も防ぐ。また、地元に酒蔵があれば、酒粕（できれば純米酒の酒粕）を入手し、表面に塗って酒粕蓋をする。これで密閉しておけばほぼカビは生えない。両方ない場合はラップで密閉する。

8 蓋をして発酵熟成させる。初めの1ヵ月ほどは、15℃以下の涼しい場所において雑菌の繁殖を抑え、まず優良な乳酸菌を繁殖させる。その後の保存場所は常温（15～30℃）でよい。1～2週間に1度は蓋を開けてチェックし、表面にカビが生えたら取り除く。

9 半年ほどおいて、表面にテカりやツヤが出て、味見して甘みや旨みがあり、香りがよく出て、塩がこなれればできあがり。美味しいと自分が感じればよい。容器から味噌を取り出し、別の容器に小分けにして密閉し、冷蔵庫などに入れてそれ以上の発酵熟成を抑えると、美味しさが長続きする。

with the masher and allow to cool.

4. When the soybeans have cooled to below 60°C/140°F, add the koji and salt mixture from step 2 and mix well. If the mixture is not liquid enough, add some cooking water to adjust (aim for a texture/softness similar to that of your earlobes). If you add cooking liquid, add more salt to balance the flavour (10% of the weight of cooking liquid added).

5. Shape the mix into balls approximately the size of your fist.

6. Throw or push the balls into the fermenting container. Try to push out all the air by pressing down hard from the top every time you add a ball.

7. Level the surface. If you have a favourite home-made miso to hand, spread a thin layer of it onto the surface. This gives the newly made miso microbes essential for fermentation, which are present in home-made miso. The layer also prevents mould growing on the surface. Alternatively, if you can get hold of some fresh sake lees (preferably junmai sake) from a sake brewery, you can spread it on the surface in the same way. Sealing the surface with alcoholic sake lees is also effective in preventing unwanted mould from growing.

8. Put on the lid and leave the miso to ferment. Put the container in a cool place (under 15°C/59°F) for the first month to prevent unwanted microbial activity. This temperature also facilitates the activity of favourable lactic acid bacteria. After the first month, miso can be stored at any temperature between 15-30°C/59-86°F. Examine the surface once every few weeks and remove any surface mould that might have grown.

9. Leave the miso to ferment for about 6 months. The miso is ready when its surface turns slightly glossy and it starts to release characteristic miso aromas. You should also taste sweet, umami-rich flavours, and the sharp saltiness from when it was made will have disappeared. Remember, 6 months is only a guideline and you can leave it as long as you want until you like the taste. When the miso is just as you like it, stop the fermentation to keep it that way. Take the miso out of the fermenting container, transfer it to an air-tight container and keep it in the fridge.

菩提酛
ぼだいもと

どぶろく

Bodaimoto / Doburoku

つくり方

How to make

菩提酛とは?

　菩提酛とは、約500年前のどぶろくのつくり方です。現在の日本酒の原型になった古い仕込み方でもあり、どぶろくとして、日本中でこの変形や類似の方法が伝承されています。菩提酛の特徴は、最初に水に乳酸菌を自然培養することです。この水を仕込み水に使うことにより、腐敗を防ぎ、安全な醸造を行なうわけです。昔の人が考えた「いかに安全に、確実に、美味しい酒をつくるか?」の試行錯誤から生まれた方法です。

What is Bodaimoto ?

Bodaimoto is a traditional sake brewing method dating back 500 years, regarded as the basis of modern sake production. A number of variations exist in different regions. It starts by cultivating naturally present lactic bacteria in water, which is then used in the mash to ensure safe brewing without bacterial spoilage. This amazing system was developed through the continuous experiments of our ancestors, who brewed delicious sake consistently and safely without the benefit of modern scientific knowledge.

　材料は米、ご飯、水、麹です。割合は米:ご飯:水:麹＝4:1:10:5です。ほかにもいろいろな割合や方法がありますが、これは家庭でつくる場合の平均的な割合です。道具は、保存容器、ご飯を包む木綿の布が必要です。

Bodaimoto has four ingredients: uncooked rice, cooked rice, water and koji. The standard ratio is 4:1:10:5, although there are many variations. You will need two containers and a cotton cloth to wrap the cooked rice.

そやし水つくり

　まず保存容器にきれいに洗った米を入れます。その上に水を加え、そこに、木綿布に包んだご飯を入れます。そして手でよくもんで、ご飯のノリ分を水に溶かし出します。その後、5〜7日間、室温におき、毎朝1日1回、このご飯をもみます。するとだんだん水に乳酸菌が繁殖してきます。表面には軽く泡が浮き、うっすら白濁し、香りはヨーグルトのホエーのような乳酸香、味は弱酸性のヨーグルトのような酸味がします。水にはっきりとこのような感応的なサインが出れば乳酸菌が繁殖したという証拠、仕込み水の完成です。この乳酸発酵した水を「そやし水」といいます。

どぶろくの仕込み

　仕込んだそやし水を、米とご飯とそやし水に分けます（そやし水は捨てないで!）。米は40〜60分蒸します。蒸している間、そやし水はどぶろく仕込み用の保存容器に入れます。そこに別につくっておいた麹を入れ、合わせてよく混ぜます。これを「水麹」と呼びます。蒸し米を入れる前に、仕込み水と麹を合わせて、水中に麹の酵素を溶かし出すわけです。そして蒸しあがるのをしばらく待ちます。

　蒸し米が軟らかく蒸しあがったら、取り出して冷まします。手で触れるくらい（40℃前後）になれば大丈夫。仕込み容器に蒸し米を入れ、全体をよく混ぜ合わせます。混ぜたらその上に、そやし水に浸けていたご飯（ヨーグルト状になっているか、そ

Preparing *soyashi mizu* (water for mash)

Wash the uncooked rice, put it in a container and add the water. Wrap the cooked rice in a cotton cloth and put it in the water with the uncooked rice. Gently massage the cloth to dissolve the starch in the cooked rice in the water. Leave the container at room temperature for 5–7 days, massaging the cloth bag once a day. Lactic acid-producing bacteria will start to multiply, and the water will become slightly cloudy and small bubbles will form on the surface. You will start to notice milky lactic acid aromas, and a mildly acidic yoghurt-like flavour. This means that the lactic acid bacteria are successfully multiplying. This lacto-fermented water is called *soyashimizu* and will be used for mashing later.

Brewing doburoku

Separate the *soyashimizu* from the uncooked and cooked rice. Don't throw away the water! Steam the uncooked rice for 40-60 minutes, and while steaming pour the *soyashimizu* into the container to be used to brew the doburoku. Put the koji into the *soyashimizu* and mix well. This mixture is called *"mizukoji"*. Soaking the koji in the *soyashimizu* before adding the steamed rice extracts water-soluble enzymes from the koji into the water. Wait until the rice has finished steaming. When the rice is soft, take it out of the steamer and let it cool until you can touch it with bare hands (around 40°C/104°F). Put the steamed rice into the brewing container and mix well. Then place the cooked rice used in the *soyashimizu* in a layer on top of the mash. The texture of the cooked rice should now be like porridge. You may only have a residue left in the bag if you massaged too much. Spread this cooked rice paste over the whole surface of the mash. This is important in the early stages of brewing, as the lacto-fermented

やし水つくりでもむ力が強すぎると、ご飯がすべて溶けてカスだけのときもある）を、その一番上の表面にのせます。ペースト状になっているので、塗るようにして全体をカバーします。この乳酸発酵したご飯が、どぶろくを最初の雑菌から守ってくれる乳酸バリアになります。それを常温で1日おきます。次の日、これを木の棒などで底までよく混ぜて、ご飯と麹をつぶし、「溶け」を促進させます。

その後、3〜7日間、毎日撹拌して発酵を促します。だんだん米が溶けて甘い香りがし、泡が湧いて発酵が元気になってきます。毎日混ぜながら、味見をして自分が好みの風味のときにそのまますくって飲んでいきます。湧きあがった泡が落ち着いた頃には、もう充分にアルコールが出ているでしょう。

菩提酛の面白さは、ゼロから命が生まれ成長していくさまを家庭サイズの瓶の中で観察できるところです。それも五感のすべてを通して、命の躍動を伝えてくれます。味・香り・見た目・音・食感。すべてが日々ダイレクトに変化し、発酵していきます。自然発酵なので、うまくいくときもうまくいかないときもあるでしょう。それでもつくり続けると、必ず発酵の神様は降りてきてくれます。

paste acts as a barrier and prevents bacterial spoilage. Leave the mash at room temperature for 24 hours. On the second day, mix and grind the mash with a wooden stick evenly from top to bottom to facilitate liquefaction.

Mix the mash every day for another 3-7 days to encourage fermentation. You will see the rice continue to dissolve during this period, and the mash will emit a sweet smell and start to bubble. These are all signs of active fermentation. Taste the mash every day, and when you like the taste it is ready to drink. Once the mash stops bubbling it should have produced enough alcohol.

It is truly fascinating to witness how life starts and evolves from nothing. Making bodaimoto allows you to closely observe the progression of life in a small container and a household setting. It doesn't require any special equipment. You can use your five senses (tasting, smelling, seeing, hearing and touching) to feel the energy of this life as conditions change every day during the fermentation process. Things may not work as the way you imagined at the beginning, but that does not mean failure. This process gives you a glimpse of wild, naturally occurring fermentation. As you patiently continue, I promise you will see a light in your fermentation journey and from then you will certainly become hooked on the world of fermentation.

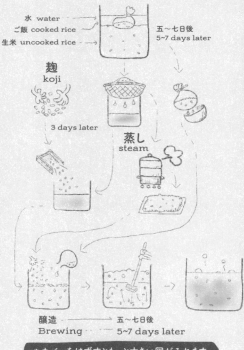

水　water
ご飯　cooked rice
生米　uncooked rice

五〜七日後
5~7 days later

麹
koji

3 days later

蒸し
steam

醸造
Brewing

五〜七日後
5~7 days later

＊カバーをはずすともっと大きい図がみれます。

Shio Koji

塩麹

塩麹は麹に塩と水を合わせて発酵させ、甘みや旨みを引き出した発酵調味料です。これまでの調味料の「味や香りを加える」という使い方ではなく（もちろんそのようにも使えますが）、塩麹の持つ酵素の分解作用で素材から味を引き出します。塩麹には麹由来の酵素、アミラーゼ、プロテアーゼが豊富に含まれています。例えば、肉や魚にまぶしてしばらくおくことで素材が分解され、甘みが増す・旨みが増す・軟らかくなる（たんぱく質が加熱凝固しない）、生臭みが消えるなどの変化がおこり、味が高まります。つまり、塩麹は酵素を使った料理法なわけです。麹の酵素をどうやって料理に生かせるか、ぜひ考えて使ってみてください。

Shio koji is an umami-rich fermented condiment with subtle sweetness, made from koji, salt and water. Unlike other condiments which add their own distinctive flavours (although shio koji can also do this), it extracts flavour from other ingredients using koji enzymes (i.e. amylase and protease). When you marinate meat or fish with shio koji, it ferments them and breaks them down, producing umami-rich, subtly sweet flavours. Proteins are not toughened by heat, so meat stays soft, yet firm. Any unpleasant fishy smells also disappear. Shio koji is a cooking method that exploits the power of enzymes. Please explore the possibility of enzymatic cooking! It is truly limitless.

How to make
➡ P54

つくり方

　材料は麹、水、塩です。割合は麹：水＝1:1に、麹と水の合計重量の6〜10%の塩です。保存容器をはかりの上にのせ、麹と水を同量で合わせます。その重さを量り、その6〜10%の好みの味で塩を入れます。それをよく混ぜ合わせておきます。その後、分解と熟成のプロセスに入ります。1日1回混ぜ、麹が完全に溶け、旨みが出るのを促します。冷蔵庫で2ヵ月熟成させるか、常温で1〜2週間おくか、熟成の方法は2通りあります。

　冷蔵庫の場合は、低温でゆっくり分解と熟成が進むので、雑菌が入りにくく、乳酸菌が主で、クリアーな味になります。常温の場合は菌数が増え、そのつくる環境によっていろんな菌が入るので、味は濃厚になり、雑多な風味が混ざります。それを味の幅や面白みとしてみることもできます。どちらでやるかは自分の暮らす環境や住んでいる国、家族の好みで決めればいいと思います。肉や魚にまぶして焼いたり、野菜にまぶして漬物にしたりします。炒め物や煮物などの料理に旨み調味料として加えてもいいでしょう。

How to make Shio koji

Shio koji has three ingredients: koji, water and salt. The ratio of koji to water is 1:1. Mix the koji and water, then add salt (6–10% of the total weight of koji and water). The easiest way to do this is to put a container on a scale and add koji and water in a 1:1 ratio. Check the weight and add 6–10% of that weight in salt. Mix the ingredients well and the fermentation process will start. Mix once a day until the koji has dissolved and umami flavour develops. You can either ferment it in the fridge for 2 month or at room temperature for about 1–2 weeks.

In the fridge, low temperature means fermentation proceeds slowly. Lactic acid bacteria dominate and prevent bacterial contamination, producing shio koji with a purer taste. A wider range of bacterial activity occurs at room temperature, resulting in richer flavours. Some people prefer these deeper flavours while others prefer the purer taste. There may also be some environmental factors involved. Choose whichever recipe you prefer. You can use shio koji for marinating meat or fish, pickling vegetables or as an umami paste for frying or stews.

麹パウダー
Koji powder

麹パウダーは、麹を乾燥させて粉末にしたものです。完全に乾燥した麹は、密閉して冷蔵庫などで保存することで、いつでも使える日常的な酵素剤になります。例えば、肉にもみ込んで軟らかくしたり、魚にまぶしたり、パンをつくるときにほんの少し加え、しっとり感を出して軟らかさを増したり。麹の粉末はアミラーゼとプロテアーゼの塊なので、食生活のさまざまなものへ応用ができるでしょう。

Koji powder is made by milling dry koji. Store it in a sealed bag and use as an enzyme powder for your everyday cooking. Rub it into meat to tenderise, or sprinkle over fish to increase umami and get rid of unpleasant fishy smells. When baking, replace some of the flour with koji powder to improve proofing. Koji powder has a wide range of uses in cooking thanks to its enzymatic profile.

つくり方

できあがった麹を広げて乾燥させます。その後、あれば野菜用の乾燥機、または40〜45℃の低温のオーブンで12時間乾燥させます。カラカラに乾燥したらミキサーやミルで粉末にします。密閉袋か瓶に入れて冷蔵庫で保存します。乾燥して密閉すれば冷蔵庫で半年以上は持ちます。

How to make Koji powder

After dekoji, spread the koji on a cloth to dry then dehydrate it in a food dehydrator or oven (keep the koji at 40–45°C (104–113°F) for 12 hours). Powder the koji using a grinder or mixer and store the powder in a sealed bag in the fridge for 6 months or longer.

Appendix-
Koji making

In this appendix, I want to introduce a method of harvesting koji spores from wild *Aspergillus oryzae* and propagating koji spores. Pedigreed spores from koji suppliers give you consistent results, which is not the case for home-cultivated koji spores. However, this method of wild cultivation has a long history and is still used in some rural areas.

Both traditional, home-cultivated koji spores and commercially marketed koji spores have strong cultural roots and should be valued. I believe we should respect both and pass them down to future generations.

もうひとつの麹つくり

ここで紹介するのは、自然の中にある麹菌やそれらを種継ぎして
培養していく方法です。ただ、家庭で育てた菌でつくる麹は、
種麹メーカーの菌とは違い、できあがりは安定しません。
それでも、自然の中から麹菌を培養する方法が昔からあり、
今も民間でごくわずかですが、行なわれています。
市販の菌と自然培養の菌、それぞれリスペクトし、
未来につなげていく必要があるとぼくは考えています。

田んぼの稲麹（いなこうじ）から育てる麹

稲麹は、稲につくカビです。日本の稲の収穫期、秋に入る頃、田んぼの稲の穂実に、濃緑色の爪の大きさほどのカビ玉が現れます。これは、稲こうじ病とも呼ばれ、この原因菌は「Ustilaginoidea virens／ウスチラジノイデア・ヴィレンス」と呼ばれます。ニホンコウジカビである「Aspergillus oryzae／アスペルギルス・オリゼー」とは違う菌です。しかし昔から日本では、この稲麹からも麹をつくってきた文化があります。

稲麹は多く現れる年と現れない年があり、昔の農村では、稲麹は豊作のサインとされていました。稲麹は稲玉、稲魂、稲霊／（いなだま）とも書きます。

稲の精霊や魂という意味も含まれていました。実際、ある日突然、田んぼに姿を現す稲麹はまさに田んぼの精霊のようです。しかもそのカビを培養すると、米が酒になるという現象が起こるのです。「Wao!!」です。そして世界中のどの文化でも、主食穀物と酒は宗教や精神文化と密接につながって発展しています。昔の日本では、稲を育てるとその収穫した米を酒にする麹菌ももれなくついていたのです。ここに自然界の大きな不思議と愛を感じるのは、ぼくだけでしょうか?

発酵醸造学分野での農学博士で東京農大名誉教授の小泉武夫氏の研究では、この稲麹の中に5〜6

種類のカビ菌が混在しコロニーになっていて、その中にオリゼーもいることがわかっています。そこで、日本の麹菌は稲麹から（または稲との共生環境の空気中から）由来したのではないか？とする説があります。また、オリゼーの祖先は毒素を持ったアスペルギルス・フラバスという野生の麹菌で、日本人に飼い慣らされるうちに毒を生産するDNAがなくなり、今のオリゼーになったとする説もありましたが、2019年12月時点では、毒素を持つアスペルギルス・フラバスとアスペルギルス・オリゼーは、遺伝的に全く別の系統であるという新しい研究結果が報告がされていま

す。麹菌のオリジナルはそもそもどこにいて、どうやって日本人と共に生きるようになったのかは、現在でもはっきりわかっていません。日本の食文化の根幹でもある麹菌の源流は、実は謎に包まれたままです。

野生のアスペルギルス属のカビには毒性を持ったものもいるので、稲麹から培養した麹を使った食品を他人に提供する場合は、専門機関の検査を受けるのが望ましいでしょう。もしあなたに知的好奇心と麹への情熱があるなら、ぜひチャレンジしてみてください。新しい発見もあるかもしれません。

Cultivating koji spores from inakoji

Inakoji is a collection of symbiotic moulds that live on rice plants. They are small (roughly the size of your nail), dark green and spherical. They commonly appear on the tip of rice ears at the beginning of autumn, during harvest season. They are a sign of "false rice smut", a rice disease caused by Ustilaginoidea virens, which is a different species from Aspergillus oryzae, the Japanese koji mould. However, our ancestors also cultivated koji from inakoji.

Inakoji appears during harvest in some years and not in others, and was therefore believed to be a sign of a good crop. It appears spontaneously in rice fields and can be used to brew sake. Therefore, it is not surprising that our ancestors believed it to be an embodiment of the divine. Staple grains and alcoholic drinks are deeply associated with religion and spirituality in many cultures. In Japan, rice was grown and koji followed. Rice and koji together make sake. I still feel awe at the marvels and bounty of nature, which is far beyond coincidence.

A research paper published by Prof. Takeo Koizumi, an honorary professor at Tokyo University of Agriculture and leading agricultural expert in fermentation and brewing, identified colonies of approximately 5 different species

of microbes in inakoji, one of which is oryzae. From this finding, a hypothesis was developed in which Japanese koji may have originated from inakoji (or from the rice ear it grows on). Another hypothesis is that Aspergillus oryzae originated from another wild member of the Aspergillus family called flavus, which has genes for producing toxins. These genes seem to have been selectively removed throughout the long history of koji cultivation. However, new findings published in December 2019 argue against this hypothesis, and state that oryzae and flavus are not genetically related. As of now, where, when and how oryzae came into the lives of the Japanese has yet to be made clear. The origins of koji, so central to Japanese food culture, are still veiled in mystery.

Some strains of wild Aspergillus mould produce toxins. From a safety point of view, it is strongly recommended to have koji analysed by specialist laboratories before using it for food production if you cultivate it from inakoji. If you are lucky enough to get hold of inakoji and your curiosity and koji passion are stimulated, why don't you try growing koji spores from wild Aspergillus oryzae? It may open a new dimension of koji making for you.

つくり方

材料…一～七分搗き米、稲麹（稲玉）（米の3％）、椿灰（米の3％）

　稲麹は、晩生の品種や、自然農や無農薬の田ん

ぼに出やすいといわれています。穂についた黒いカビの玉が稲麹で、そのような田んぼで稲の出穂、穂が出た後から収穫前の8～10月に見られます。稲麹がついた穂の先端ごと摘み取り、穂につけたまま風

に当てず、よく乾燥させます。箸などでしごいて、カビ玉の部分だけを使います。稲麹はできるだけ新鮮なものがオススメですが、保存もできます。穂のまま乾燥させたら、木灰（できればきれいな椿灰）をまぶし、密閉袋や瓶に入れて保存しておけば、1年以上たっても使えます。

米は本来、玄米から外側1〜2％だけ削った98％精米を使いますが、一般家庭では難しいので、ここでは手に入りやすい七分搗きの米を使います。椿灰は染色材料、陶芸材料の店、椿専門の園芸店などで手に入ります。純粋に椿灰のみのものを用意してください。

米は、通常の麹つくりのように浸漬、水切りして蒸します。軟らかく蒸しあがったら、滅菌したバットに取ります。種麹を培養するときは、素手で触ることは極力避け、米に触れる道具は全て熱湯消毒し、よく乾かして清潔に保っておきます。

米が熱いうちに（65℃以上で）椿灰をまぶしかけます。熱いうちに灰をかけるのは、雑菌が蒸し米に付着する前に、アルカリ性の環境にして雑菌の繁殖を抑制するためです。その後、灰が全体に行き渡るように木ベラやスプーンなどで混ぜ続けて、蒸し米が45℃になったら稲麹の胞子を振りかけます。茶こしや目の細かいザルに入れて振るとよいでしょう。全体にまんべんなく振れたらよく混ぜます。蒸し米が34〜36℃でこの作業を終わるようにします。その後、通常の包み込みと同じようにバットに広げて、温度計を差し、もう1枚のバットで蓋をして、全体を電気毛布や保温機で、34〜38℃をキープします。湿度は90％以上を目指します。

保温と手入れも通常の麹つくりと同じように1〜3日目までの作業をします。ただし3日目の最高温度は38℃までです。仕舞仕事の後は、38℃からゆっくりと36℃まで下げて、その後5〜7日間保温保湿を続けます。3日目以降はバットだけでは保湿できないので、湿らせた布をバットにかぶせて蓋をします。布は1日1〜2回取り替えます。これは最後の8〜10日目まで行ないます。

4日目以降から、徐々に麹の表面に胞子ができ始め、そのうちに黄色から緑に変化し、全体が緑色の胞子で覆われます。5〜6日目には胞子の量がどんどん増え始め、7〜8日目にはびっしりと緑の胞子で覆われます。充分に胞子が成長したらできあがり。出麹とします。これを風のないところで充分に乾かし、2〜3日枯らし、その後、胞子が飛ばないようにやさしく密閉袋に入れ、冷暗所や冷蔵庫に保管します。通常の麹つくりと同じように、この緑の胞子を種麹として、蒸し米に振りかけて使います。もし、市販する食品に使用するときには、必ず微生物専門の検査機関に検査してもらい、安全を確認してから使用してください。微生物を扱うときは、ロマンと安全を両立させる責任が製造者にはあります。

How to make koji spores from inakoji

Ingredient: 10-70% polished rice, ina koji (3% of rice weight), camellia ash (3% of rice weight)

Inakoji is often found on late-growing rice varieties or organically or naturally grown rice. Small dark green spheres are seen at the tip of the ears of rice from August to October, when the ears start appearing and maturing. Cut the inakoji off along with the ears and dry them by placing them in a room without air currents. Strip off the spheres with chopsticks (you use only the spheres to grow koji). The success rate is higher with fresh inakoji, but you can also keep it in a sealed bag for over a year if properly dried. In this case, keep them on the ears and mix them with the ash (preferably Camellia).

Brown rice with 1-2% polished off is traditionally used as a substrate. This rice is hard to find, so you can also use rice bought from rice shops with 70% threshed off. You can find camellia ash in shops selling dye products, pottery shops or specialised garden shops selling camellia. Use pure camellia ash.

Soak, drain and steam the rice as you would do when you make ordinary koji. Place the steamed rice on a sterilised tray. To avoid bacterial contamination, try not to touch the rice with your bare hands. Use boiled water to sterilise all equipment that will be in contact with the rice, and dry everything well.

Sprinkle the camellia ash while the rice is still hot (over 65°C/149°F) to create alkaline conditions before it comes in contact with other bacteria. Mix the rice and ash well with a sterile wooden spatula or stainless steel spoon. Wait until the rice has cooled down to 45°C (113°F), sprinkle the spores from the inakoji evenly using a tea strainer or fine sieve, and mix well. Move onto the next step while the rice is around 34-36°C (93-97°F). Spread the rice on the tray as you would normally, insert the thermometer and close the lid using another tray placed upside down. Keep the rice between 34-38°C (93-100°F) using an electric blanket or other warming equipment. Optimal humidity is over 90%. Follow the usual 3-day process to make koji, trying not to exceed 38°C (100°F). After the third mixing (shimai), lower the temperature gradually from 38°C (100°F) to 36°C (97°F) and maintain the temperature and humidity for another 5-7 days. After the 3rd day, place a moist cloth over the tray to keep everything humid. Replace the cloth a few times a day. Repeat until the 8th to 10th day.

Spores start appearing from the 4th day onwards and their colour will gradually change from yellow to green. Spores will visibly increase from around the 5th day and the entire grain will be covered with green spores by around the 7th day. Continue the process until enough spores are produced, then take the koji out of the tray (dekoji). Dry the spores for a few days in a room without air currents. Transfer the grains very gently into a bag, ensuring you do not lose any spores, and seal the bag. Keep the bag in the fridge or in a dark, cool place. The grains can be used in the same way as commercial koji spores, by sprinkling over steamed rice. It is vital to have the koji examined by specialist microbial laboratories before using it for food. When you pursue your passion, keep in mind that it is your responsibility to ensure safety using scientific measures.

麹からの種を継ぐ　～種麹をつくる

　種麹が入手しにくい国にお住まいの方、インターネットにアクセスしにくい方へ。

　日本国内でインターネットができれば、今はいつでもどこでも種麹は入手できます。でも、海外やインターネットにアクセスしにくい環境にいる方もいるでしょう。そんな方のために、ここでは参考として、できた麹から種を継ぐ、種麹をつくる方法を紹介します。

　麹つくりの「出麹」（p38）のタイミングを5～7日間引き延ばし、3日目以降さらに保温と保湿を続けると胞子がどんどん成長していきます。胞子の色は濃くなり、数も増えていきます。胞子形成の最適温度は34～36℃です。あまりに胞子が成長すると麹としては使いにくいのですが、これは胞子なので種麹として使えます。

　こうした麹から胞子をとって種を継ぐことは、もともとの日本の農村、味噌蔵や醤油蔵でも行なわれていました。西欧で自家製のサワードウ（パン種）やチーズの菌を継いでいくように、麹も昔はおおらかに種菌を継いでいたのです。

　種麹をつくるときには、通常の麹つくりとは違う認識で麹を育てます。通常の麹つくりは「酵素を生産する」ことが目的ですが、種麹つくりは、「胞子を生産する」ことが目的です。同時に衛生面に気を配り、雑菌の混入を極力防がなければなりません。普通に麹をつくり、その一部を種にするという方法では、すぐに雑菌だらけの種麹になってしまいます。

　容器や道具は完全に熱湯殺菌したものを用意し、初めの蒸し米が60℃以上の熱いときに原料の3％以上の「木灰」を加えて強アルカリ環境の培地をつくり、種切りをして、温度34～36℃、湿度70～90％の環境を維持して8～10日間、麹の培養を続けます。そこに次世代の胞子がつき、それを無風状態で乾燥させればその胞子を種麹として使えます。完全に乾燥させて密閉袋に入れれば1年以上、冷凍すればさらに長く保存できます。カビの胞子は、基本的には植物の種子のような扱いで保存ができます。

　その胞子を蒸し米に振りかければ、また麹つくりを始められます。ただこの胞子を種継ぎし続けることは、基本的にはオススメしません。なぜなら世代交代を経るうちにどんどん本来の性質は変わっていき、野生のカビ菌とも混ざっていくからです。すると常にその菌が安全かどうか？という点に注意を払う必要が出てきます。家庭で種継ぎしたものは、基本的には最初の種とは全く違うものになっていることを認識しておきましょう。だから、できあがった酒や味噌の味や香りが変わってきたときには種麹メーカーのつくる、純粋な種麹を新たに使うことをオススメします。メーカーの種麹を使ったほうが、食品の安全と品質は維持しやすいです。

　暮らしの中で育まれた発酵文化を知るという知識と技術の継承という側面では、種継ぎのことも知っておいたほうがよいと僕は考えています。それも踏まえつつ、種麹を純粋に500年以上も守り伝えて、発酵醸造文化を支えてきた種麹メーカーをリスペクトしてい

ます。そして味や酵素量、その品質の安定、安全面
はやはり種麹メーカーのほうが優れています。農村
部で伝承されてきた民間の麹つくり、蔵の中で専門

家の技術として発達・洗練されてきた醸造業界の麹
つくり。その二つは相互に補完しあい、二つとも日本
の発酵文化において大切な宝物です。

Tanetsugi
Propagating another generation of koji spores

Koji spores can be easily purchased online in Japan.
However, if you are outside Japan or have no internet
access, they may be difficult to get hold of. The good
news is that when you make koji you can also create new
generations of koji spores to use in another batch. This
process is called *tanetsugi*.

If you delay *dekoji* (p.38) for a few days and maintain
the right heat and humidity, more and more spores will
be produced and the grains become darker in colour. The
optimum temperature for this additional growing period
is 34–36°C (93–97°F). In the normal koji making process
you avoid creating spores, but the aim here is to produce as
many spores as possible.

This method of propagating spores was traditionally
practised in farming villages or breweries. Koji spores were
once propagated casually, like starter cultures for sour
dough or cheese.

You must change your focus from the usual koji making
process. I explained before that the purpose of making
koji is to produce enzymes, but the purpose here is
to produce spores. You must also take extra care with
hygiene and make even more of an effort to avoid bacterial
contamination. This is because if you simply use the usual
koji making method, unwanted bacteria will be attracted
to the substances koji produces during the longer growing
period.

Make sure to sterilise all containers and equipment using
boiled water. Prepare a strong alkaline substrate by mixing
in ash (over 3% of uncooked rice weight) before the rice
has cooled down below 60°C (140°F). Sprinkle the spores
and maintain the temperature at 34–36°C (93–97°F)
and humidity at 70–90% for the 8–10 days. Once dried
in a room with no air currents, the spores for the next

generation can be used to make koji. Properly dried koji
spores can be preserved for over a year if kept in a sealed
bag (even longer if you freeze them). You can keep and use
them in the same way as plant seeds.

You can start growing koji by sprinkling the propagated
spores over steamed rice. However, continuously
propagating spores is not recommended as their properties
may change across generations. Safety cannot be
guaranteed if the koji breeds with other wild moulds. Please
be aware that the nature of koji spores propagated at home
could be very different from the original pedigree spores.
If you notice the aroma or flavour of fermented foods
made using home-propagated spores is clearly different, I
recommend obtaining fresh spores from professional koji
producers. It is much easier to ensure safety and quality
of fermented foods if you use professionally grown koji
spores.

However, the reason I introduce the tanetsugi method in
this book is because I believe it is culturally important
to pass down the knowledge and techniques of these
traditional cultures to the next generation.

I cannot stress enough the amount of respect I have
for koji spore producers who have supported Japanese
fermentation culture by cultivating high quality spores for
over five centuries. Without doubt, the pedigree spores
are by far superior in every way: enzyme production,
consistency and safety when making koji and the flavour of
the final products. However, these are two very different
koji making techniques, which complement one another. One
has developed naturally as part of farming culture, while
the other has been developed by professional producers and
enhanced by tireless collaboration with breweries. Both are
precious treasures of Japanese fermentation culture.

本書掲載のQRコードまとめ

　今回の本では麹つくりの各動作をテキストだけでなく動画でも見てもらえるように、以下の動作の動画を用意しました。本文中に掲載されているQRコードか、以下のQRコードをスマートフォンで読み込むと、そこから動画のYouTubeページに飛ぶことができます。

　また今後、麹のつくり方や使う道具について、新しい方法やアイデアを思いついたら、こちらのなかじのYouTubeチャンネルで情報を随時更新していきます。できた麹の使い方、味噌、甘酒、料理のつくり方などの動画もアップしますのでぜひ参考にしてください。

道具 TOOLS AND EQUIPMENT
https://youtu.be/9ClfJbfhMkM

洗米 SENMAI
https://youtu.be/tmTtaClZGl8

かし KASHI
https://youtu.be/DO3ytN2coZo

蒸し MUSHI
https://youtu.be/AaZifzVPNtM

ひねりもち HINERIMOCHI
https://youtu.be/F5sRd3HiRgg

バットの滅菌 STERILISE TRAYS
https://youtu.be/nvm181kqNyk

種切り TANEKIRI
https://youtu.be/r3YV7MlqBSA

出麹と枯らし DEKOJI-KARASHI
https://youtu.be/7OlWgfth8vY

まとめページ PLAYLIST
https://www.youtube.com/playlist?list=PLMvNiNzGlIcC3hreQEJgbAcQxqL9Y2QOS

種麹の入手先一覧
List of koji spore producers
（2020年2月現在）

株式会社 秋田今野商店
AKITA KONNO CO., LTD.
http://www.akita-konno.co.jp
〒019-2112 秋田県大仙市字刈和野248
TEL 0187-75-1250（代表）
FAX 0187-75-1255

日本醸造工業／丸福
NIHON JYOZO KOGYO CO.,LTD
http://www.nihonjouzou.co.jp
〒112-0002 東京都文京区
小石川3丁目18番9号
TEL 03-3816-2951
FAX 03-3814-9666

石黒種麹店
ISHIKURO TANEKOJI TEN
http://www.1496tanekouji.com
〒939-1652 富山県南砺市福光新町54番地
TEL 0763-52-0128／FAX 0763-52-0184
問い合わせメール e-miso@amber.plala.or.jp

糀屋三左衛門／関連会社ビオック
KOJIYASANZAEMON CO., LTD.
http://www.koji-za.jp
〒441-8087 愛知県豊橋市
牟呂町内田111-1
TEL 0532-31-0311（代表）

菱六
HISHIROKU
〒605-0813 京都府京都市東山区松原通
大和大路東入二丁目 轆轤町79
TEL 075-541-4141

樋口松之助商店
HIGUCHI MATSUNOSUKE SHOTEN CO., LTD.
http://www.higuchi-m.co.jp
〒545-0022 大阪府大阪市
阿倍野区播磨町1丁目14番2号
TEL 06-6621-8781
FAX 06-6621-2550

河内源一郎商店
KAWAUCHI GENICHIRO SHOTEN
https://www.kawauchi.co.jp
〒899-6404 鹿児島県霧島市
溝辺町麓876-15
TEL 0120-37-0995
FAX 0120-58-3157

AFTERWORD

おわりに

麹と感性

　今、読んだこの本と、あなたの顔との空間には、夜空の星の数以上の小さな微生物が浮遊しています。そして、ヒトの体表面や体腔の内側にも微生物はひしめき、ぼくたちは常に無数の見えない命に包まれて生きています。麹つくりは、そのような小さな命の躍動を、五感で感じ取ろうとする感性のトレーニングです。

麹と環境

　ヒトは産まれてから死ぬまで、空間的に多くの微生物に包まれ、機能的には協力し、微生物と共に生きてきました。このようなヒトと微生物の共生関係と、地球全体の生態系には多くの共通項があります。どちらも健康的に持続していくには多様性が必要です。多様性という空間の彩りが、持続性という時間の繋がりをつくっています。ぼくたちがこの地球で、これからも健康的に生きて行くために、微生物から学び、微生物と共に生きていく視点が、これからの時代を生きていくのに大きなヒントをくれるのでは?と思います。

麹と社会

　キッチンで麹を育てながら、麹の成長する様子に生命の循環をイメージしてみてください。この中に小さな宇宙があります。そんな視点で見てみると微生物から多くのことを学べるのでは?と思います。ぼくたちが、微生物を含めた多くの生き物と共存して生きている、という体感を伴う体験が広がることは、社会をより発酵的にしていくでしょう。その流れはキッチンから始まり、暮らし方全般へ。そして農業、健康、教育、あらゆる社会全体のシステムへ広がっていくと考えています。麹の技術と知恵を広げ、このような発酵体験を共有したいと思っています。いつか麹の世界でお会いしましょう。Love & Oryzae!

Koji and your senses

Billions of tiny microbes exist in the space between you and this book you are holding. There are more microbes than stars in the sky. Huge numbers live on and inside your body, demonstrating how we are constantly surrounded by countless living things. By making koji, you train your five senses to feel the vibrant energy of these minute creatures.

Koji and the environment

From the very moment of birth, throughout life and until death, humans are surrounded by infinite numbers of microbes that harmoniously cooperate and live together with us. There are many similarities between the symbiotic relationship between humans and microbes, and the ecological system that sustains the entire globe. Both require adaptability and diversity for healthy development. In other words, sustainable evolution is only possible by preserving and protecting the richness of biodiversity. We need to borrow the wisdom of microbes and make an effort to create a mutually beneficial living environment in order to carry on leading sound lives on this beautiful planet.

Koji and society

When making koji in your kitchen, try to imagine its growth as the circle of life. You will start to see the universe these tiny creatures live in. Changing your point of view reveals completely different scenes and a new perspective. In this sense, I think we can learn a great deal by projecting ourselves onto the microbial world. Once you realise how our lives are supported by the harmonious coexistence and cooperation of many creatures, and have this understanding proven through your own experience, it can change your life. A small experiment started in the kitchen can gradually affect the way you think, your relationships with others, and eventually your whole world. If more people experienced this, it could lead to a movement powerful enough to change society. It could influence a whole range of social concerns, such as agriculture, health and education. I truly believe there is great potential in this "fermentative" experience. By sharing my skills and knowledge of koji, I hope to share this experience with as many people as possible. I hope to see you in the koji universe. Love & oryzae!

なかじ

麹文化研究家。麹の学校主宰。料理研究家中島デコ氏に師事。造り酒屋「寺田本家」で酒造りに携わり蔵人頭となる。現在は全国で発酵のセミナーと、麹のワークショップを開催。日本及び海外で麹の学校を開催し、麹の文化と技術を伝えている。オンラインで麹を学ぶ「麹の学校サロン」を主宰。著書に『酒粕のおいしいレシピ』（農文協）ほか。HP/SNS…https://www.nakaji-minami.com/ YouTube…なかじnakaji's fermentation journey
■麹を学ぶオンライン・コミュニティ「麹の学校サロン」で検索。

写真 ─────── 寺澤太郎
デザイン ─── 藤田康平（Barber）
翻訳 ─────── Haruko Uchishiba
協力 ─────── Arline Lyons
　　　　　　　Saki Uchishiba
　　　　　　　Emi Uchishiba
　　　　　　　Juri Uchishiba

麹本
KOJI for LIFE

2020年 3 月25日　第 1 刷発行
2024年10月25日　第 6 刷発行

著者 ─────── なかじ

発行所 ─────── 一般社団法人 農山漁村文化協会
　　　　　　　〒335-0022　埼玉県戸田市上戸田2-2-2
　　　　　　　電話 048-233-9351（営業）
　　　　　　　電話 048-233-9372（編集）
　　　　　　　FAX 048-299-2812
　　　　　　　振替 00120-3-144478

〈検印廃止〉
ISBN 978-4-540-19131-2 定価はカバーに表示
©Nakaji 2020 Printed in Japan

印刷・製本 ─ TOPPANクロレ株式会社

乱丁・落丁本はお取り替えいたします。